Understanding Smart Sensors

Artech House Related Titles

Industrial Microwave Sensors, Ebbe G. Nyfors and Pertti V. Vainikainen

Introduction to Sensor Systems, Shahen A. Hovanessian

Optical Fiber Sensors, Volume I: Principles and Components, John Dakin and Brian Culshaw

Optical Fiber Sensors, Volume II: Systems and Applications, John Dakin and Brian Culshaw

Sensor Technology and Devices, Ljubisa Ristic

Understanding Smart Sensors, Randy Frank

For further information on these and other Artech House titles, contact:

Artech House
685 Canton Street
Norwood, MA 02062
617-769-9750
Fax: 617-769-6334
Telex: 951-659
e-mail: artech@world.std.com

Artech House
Portland House, Stag Place
London SW1E 5XA England
+44 (0) 171-973-8077
Fax: +44 (0) 171-630-0166
Telex: 951-659
e-mail: bookco@artech.demon.co.uk

Understanding Smart Sensors

Randy Frank

Artech House
Boston • London

Library of Congress Cataloging-in-Publication Data
Frank, Randy.
    Understanding smart sensors/Randy Frank.
        p. cm.
    Includes bibliographical references and index.
    ISBN 0-89006-824-0 (alk. paper)
    1. Detectors—Design and construction. 2. Programmable controllers. 3. Signal processing—
Digital techniques. 4. Semiconductors. 5. Application specific integrated circuits.
I. Title
TA165.F724 1996   95-48912
681'.2–dc20   CIP

A catalogue record for this book is available from the British Library

© 1996 ARTECH HOUSE, INC.
685 Canton Street
Norwood, MA 02062

International Standard Book Number: 0-89006-824-0
Library of Congress Catalog Card Number: 95-48912

10 9 8 7 6 5 4 3 2 1

Dedicated to the memory of the one person who would have loved to see this book but did not—my father, Carl Robert Frank

Contents

Preface

By the year 2000, 50% of all engineers will design with sensors, up from 16% who routinely used them at the beginning of the decade (Gardner, D. L., "Accelerometers for Exotic Designs," *Design News*, July 17, 1989, p. 55). Micromachining technology will be the primary reason for sensors achieving cost breakthroughs that allow this widespread sensor usage. At the heart of most smart sensors will be digital electronic control.

Embedded microcontrollers already play a hidden role in most of the common activities that occur in our daily lives. Use a cellular phone, receive a page, watch television, listen to a compact disc, or drive a current model car and you have the assistance of embedded microcontrollers. For example, there are 10 microcontrollers in a typical car, or over 50 if you drive a luxury car. Many of the inputs to these devices are provided by semiconductor sensors. The number of sensors and the intelligence level is increasing to keep up with increasing control complexity.

Semiconductor sensors were initially developed to provide easier-to-interface, lower cost, and more reliable inputs to electronic control systems. The microcontrollers at the heart of these systems have increased in complexity and capability while achieving drastically reduced cost per function. Semiconductor technology has also been applied to the input side for a few sensor inputs (pressure, temperature, acceleration, optoelectronics, and Hall effect devices), but is just starting to broaden in scope (level of integration) and sensed parameters and achieve cost-reduction benefits from integration.

The system outputs have done a better job of keeping up with advances in semiconductor technology. The term "smart power" refers to semiconductor power technologies that combine an output power device(s) with control circuitry on the same silicon chip. Both input and output devices are receiving greater focus, the capability of combining technologies is being extended, and the need for systems-level communications is finally making smart sensors a reality.

Joe Giachino of Ford Motor Company is frequently given credit for the term smart sensor based on his 1986 paper (Giachino, J. M., "Smart Sensors," *Sensors and Actuators,*

10 (1986), pp. 239–248). However, several others (Middelhoek, Brignell, etc.) will claim part of the credit for pioneering the concept of a sensor with capabilities beyond simple signal conditioning. (I think I used the term in this context at least once before 1986). The communication of sensory information is finally requiring concensus for the true meaning of smart sensor.

The ultimate capabilities of new smart sensors will undoubtedly go far beyond today's projections. An understanding of what is possible today and what can be expected in the near future is necessary to take the first step toward smarter sensing systems. This book is intended to provide the reader with knowledge regarding a broad spectrum of possibilities based on the current R&D efforts of industry, universities, and national laboratories. It discusses many recent developments that will impact sensing technology and future products.

I would like to extend my sincere appreciation to Mark Shaw, whose concept of the phases of integration became an underlying theme for this book and (I believe) the way that smart sensing will evolve. A number of other people played an important role in making this book a reality.

Ray Weiss of *Computer Design* magazine provided methodology guidance and was the prime mover.

Mark Walsh and the team at Artech were very supportive at every step in the process.

Lj. Ristic, Mark Shaw, Cindy Wood, and Mark Reinhard from Motorola Semiconductor Products Sector provided chapter reviews.

ImageLab and Case Western Reserve University provided items for the cover artwork.

Dorothy Rosa and the folks at *Sensors* Magazine and *Sensors Expo* provided many publishing opportunities that were helpful in documenting several aspects of smart sensors.

However, this book would not have been possible without the critical evaluation, tolerance, and encouragement of my wife Rose Ann. She now knows how small the appreciation is when compared to the sacrifices that she made.

Chapter 1
Smart Sensor Basics

"A rose by any other name would smell as sweet."

—William Shakespeare

"A rose with a microcontroller would be a smart rose."

—Randy Frank

1.1 INTRODUCTION

Just about everything today in the technology area is a candidate for having a "smart" prefix added to it. The term "smart sensor" was coined in the mid 1980s and since that time several devices have been called smart sensors. The intelligence required by these devices is available from microcontroller (MCU), digital signal processor (DSP), or application-specific integrated circuit (ASIC) technologies that have been developed by several semiconductor manufacturers. Some of these same semiconductor manufacturers are actively working on smarter silicon devices for the input and the output side of the control system as well. To understand what is occurring today when advanced microelectronic technology is applied to sensors, a brief review of the transitions that have occurred is in order.

Before the availability of microelectronics, the sensors or transducers that were used to measure physical quantities, such as temperature, pressure, or flow, were usually directly coupled with a readout device, typically a meter that was read by an observer. The transducer converted the physical quantity being measured to a displacement. System corrections were initiated by the observer to change the reading closer to a desired value, as shown in Figure 1.1 [1].

Today many home thermostats, tire pressure gauges, and factory flow meters still operate in this same manner. However, the advent of microprocessor technology initiated the requirement for sensors to have an electrical output that could be more readily interfaced to provide unattended measurement and control. This also required the analog signal level to be amplified and converted to digital format prior to being supplied to the process controller. Today's MCUs and analog to digital (A/D) converters typically have a 5-volt

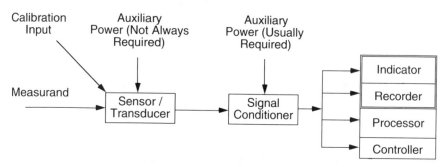

Figure 1.1 Original sensing systems.

power supply, which has dictated the supply voltage for many amplified and signal conditioned sensors.

Commonly used definitions for sensors and transducers must be the first in the list of many terms that will be defined either in this or the remaining chapters, or in the glossary that follows. A *transducer* is a device converting energy from one domain into another, calibrated to minimize the errors in the conversion process [2]. A *sensor* is a device that provides a useful output to a specified measurand. The sensor is a basic element of a transducer, but may also refer to a detection of voltage or current in the electrical regime that does not require conversion. Throughout the remainder of this book, the terms will be used synonymously, since energy conversion will be part of every device that is discussed. A list of possible mechanical measurements that require a transducer to provide an electrical output is shown in Table 1.1.

Table 1.1 Mechanical Measurements

Measurement	*Typical/Common Techniques*
Displacement/position	Variable reluctance, Hall effect, optoelectronic
Temperature	Thermistor, transistor Vbe
Pressure	Piezoresistive, capacitive
Velocity (linear/angular)	Variable reluctance, Hall effect, optoelectronic
Acceleration	Piezoresistive, capacitive, piezoelectric
Force	Piezoresistive
Torque	Optoelectronic
Mechanical impedance	Piezoresistive
Strain	Piezoresistive
Flow	Δ pressure
Humidity	Resistive, capacitive
Proximity	Ultrasonic
Range	Radar
Liquid level	Ultrasonic
Slip	Dual torque
Imminent collision	Radar

The definition of a smart sensor (intelligent transducer) has not been as widely accepted and is subject to misuse. However, the Institute of Electrical and Electronic Engineers (IEEE) TC-9 committee has been actively consolidating terminology that applies to microelectronic sensors and has defined a *smart sensor* or *smart transducer* as a device with a built-in intelligence, whether apparent to the user or not [2]. This definition provides a starting point for the minimum content of a smart sensor. However, future smart sensors will be capable of much more and additional classifications (e.g., smart-sensor type 1) may be required to differentiate these products.

1.2 MECHANICAL-ELECTRONIC TRANSITIONS IN SENSING

An early indication of the transition from strictly mechanical sensing to electronic techniques is demonstrated in the area of temperature and position measurements. Thermocouples and expansion thermometers were replaced by thermistors and semiconductor temperature sensors that were lower in cost, smaller in size, and easier to interface to other circuit elements. In position measurements, variable reluctance measurements with magnetic pickups have been replaced by Hall effect, opto sensing, and even newer magnetoresistive (MR) elements. All of these techniques take advantage of a previous problem that detracted from the ideal performance of a transistor or integrated circuit. The sensitivity of transistors to temperature, light, magnetic fields, stress, and other physical variables is exploited in semiconductor sensors.

The expanding range of parameters that can be sensed using semiconductor technology is part of the increasing interest in smart sensing. Using micromachining or chemical etching techniques, mechanical structures have been produced in silicon that have greatly expanded the number and types of measurements that can be made. For example, rubber diaphragms connected to potentiometers can be replaced by silicon diaphragms for measuring pressure. This approach has been used in production sensors for almost two decades. More recently, cantilever beams and other suspended structures have been manufactured in silicon, and acceleration is measured by resistive, capacitive, or other techniques. Table 1.2 shows a number of sensing techniques and their status relative

Table 1.2 Sensing Techniques

Technique	Status in Silicon Sensor
Piezoresistive	Pressure, acceleration
Capacitive	Pressure, acceleration
Piezoelectric	Pressure, acceleration
Optoelectronic	Position, velocity
Magnetic	Position, velocity, magnetic field
Radar	Limited production
Lidar	Production/research/development
Ultrasonic	Production

to implementation in silicon sensors. Chapter 2 will explain the most popular developments in micromachining.

Processes used to manufacture advanced semiconductor technology are being
adapted by sensor manufacturers. As a result, sensors are being manufactured, either
concurrently or separately, that take advantage of the performance enhancements that
integrated circuit (IC) technology can provide, and a significant step forward is occurring
in sensing technology. Chapters 3 and 4 will develop these interfacing and integration
aspects.

1.3 NATURE OF SENSORS

The output from most sensing elements is low level and is subject to several signal
interference sources, as shown in the generalized model of a transducer, Figure 1.2 [3].
Self-generating transducers such as piezoelectric devices do not require a secondary input
to produce an output signal. However, transducers based on resistive, capacitive, and
inductive sensing elements require excitation to provide an output. In addition to the desired
input (for example, pressure), undesired environmental effects, such as temperature, humidity, or vibration, are factors that affect the performance and accuracy of the transducer
and must be taken into account during the design of the transducer. Compensation for these
secondary parameters has historically been performed by additional circuitry, but with
smart-sensing technology the compensation can be integrated on the sensor or accomplished in the microcontroller.

The output of a micromachined piezoresistive silicon pressure sensor and the effect
of temperature on both the span and offset is demonstrated in Figure 1.3 [4]. Although the

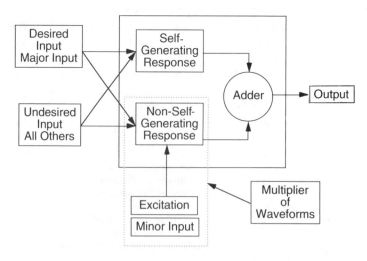

Figure 1.2 General transducer model. (*After:* [3].)

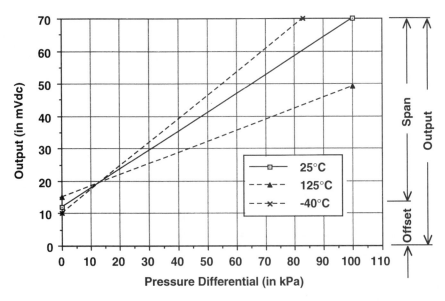

Figure 1.3 Effect of temperature on piezoresistive pressure sensor output.

output is quite linear, in this case within 0.1% full scale (F.S.), the output varies due to the effect of temperature on the span of the sensor by about 0.12 mV/°C. Since this signal level is insufficient to directly interface to a control IC, additional amplification and calibration are typically performed in the next stage of a transducer.

In a simple control system, the sensor is only one of three items required to implement a control strategy. The (1) sensor provides an input to a (2) controller with the desired strategy in its memory. The controller drives an (3) output stage to modify or maintain the status of a load, such as a light, motor, solenoid, or display. As shown in Figure 1.4, a signal-conditioning interface typically exists between the sensor(s) and the controller and between the controller and the output device. Smart sensing has meant that a portion of the controller's functions are included in the sensor portion of the system.

The smart-sensor models developed by several sources [5–7] have as many as five distinct elements for analog output sensors. As shown in Figure 1.5, in addition to the sensing element and its associated amplification and signal conditioning, an analog to digital converter, memory of some type, and logic (control) capability are included in the smart sensor. Once the signal is in digital format, it can be communicated by several communication protocols. The regulated power supply that is also required for the system and its effect on system accuracy are not taken into account in this model.

Reducing the number of discrete elements in a smart sensor, or any system, is desirable in order to reduce the number of components, form factor, interconnections, assembly cost, and frequently the component cost as well. The choices for how this

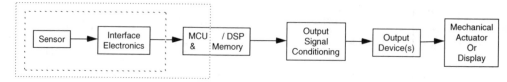

Figure 1.4 Generic control system.

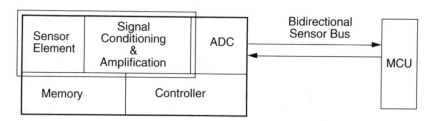

Figure 1.5 Smart-sensor model.

integration occurs are often a function of the original expertise of the integrator. For example, as shown in Figure 1.6, a sensor manufacturer that already uses semiconductor (i.e., bipolar or metal oxide semiconductor (MOS)) technology for the sensing element may expand the capability and increase the value (and intelligence) of the unit that is produced by combining the signal conditioning in the same package or in a sensor module. Through integration, the signal conditioning can also be combined at the same time the sensor is fabricated or manufactured. While the process of integration is more complex, the integrated sensor can be manufactured with the sensor and signal conditioning optimized for a particular application. Conversely, an MCU manufacturer using a complementary metal oxide semiconductor (CMOS) process typically integrates memory, A/D, and additional signal conditioning to reduce the number of components in the system. A variety of combinations is indicated in Figure 1.6. Processing technology is a key factor. However, manufacturers must not only be willing to integrate additional system components, they must also achieve a cost-effective solution. Combinations of hybrid (package level) and monolithic integration will be discussed frequently in the remainder of this book. Different design philosophies and the necessity to partition the sensor/system at different points can determine whether a smart sensor is purchased or, alternatively, designed using a sensor signal processor or other components necessary to meet the desired performance of the end product.

The integration path can have a significant effect on the ultimate level of component reduction. As shown in Figure 1.7 [8], the input (demonstrated by a pressure sensor), computing (HCMOS microcontroller), and output side (power MOS) are all increasing in the amount of monolithic integration. The choice of sensor technology, such as bipolar, can have a limiting effect on how far the integration can progress. For example, a bipolar sensor can increase the integration level by adding signal conditioning and progress to a monolithic

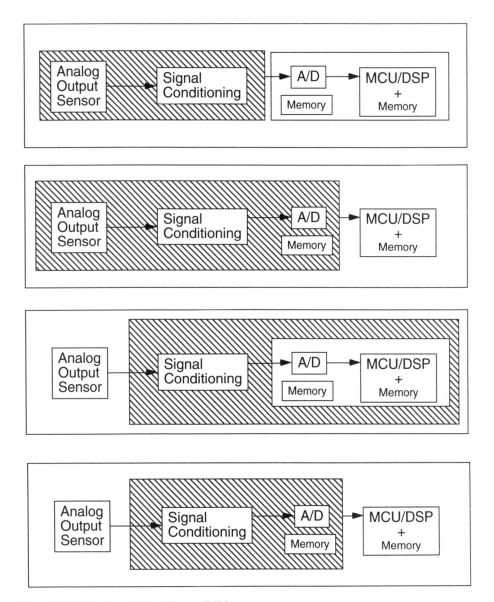

Figure 1.6 Partitioning and integration possibilities.

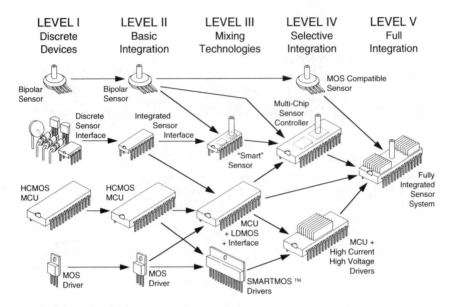

LEVEL I	LEVEL II	LEVEL III	LEVEL IV	LEVEL V
Discrete Devices	Basic Integration	Mixing Technologies	Selective Integration	Full Integration

Figure 1.7 Sensor technology migration path.

Level III sensor. Through package-level integration, a 2-chip sensor controller can be achieved by combining the sensor with a high-density CMOS (HCMOS) microcontroller. However, the highest level of monolithic integration, Level V, will only be realized by pursuing MOS-compatible sensing and power control technologies.

Realizing the full potential of these new sensors will require a new approach to identifying sensor applications. The list of "sensor" terms in Table 1.3 serves as a starting point for rethinking the possibilities for smart sensors. Many of the terms are associated with a system and not the sensor portion of the system.

When sensors are combined with an MCU, a DSP, or an ASIC, the capability to obtain additional performance improvements is limited only by the capability of the computing element and the imagination of the designer. Chapters 5, 6, and 7 explain and develop some of these possibilities. In addition to signal transmission in distributed control systems through a variety of protocols described in Chapter 6, the possibilities of portable, wireless, and remote sensing are explored in Chapter 8. A broad variety of micromechanical elements and additional system components are investigated in Chapter 9. One of the more challenging and limiting factors to higher levels of integration is packaging, and Chapter 10 reveals the progress that is being made in packaging and potential areas for future development. The combination of the previously discussed aspects is already being researched as discussed in Chapter 11. Finally, based on the system-level complexity that is already possible and continuously evolving, a look into the not-too-distant future of smart sensing is shared in Chapter 12.

Table 1.3 Terms Used to Identify
New Smart-Sensor Applications

Measure	Understand
Monitor	Diagnose
Correct	Control
Detect	Presence/absence
Safe operating area	Exceed level
Communicate	Identify
Prevent failure	Maintenance
Warning	Instrument
Regulate	Gauge
Observe	Know
Determine	See/hear/touch/smell/taste
How/what/where/when/why	

1.4 INTEGRATION OF MICROMACHINING AND MICROELECTRONICS

Increasing the performance and reliability, and reducing the cost of electronic circuits through increased integration, are standard expectations for semiconductor technology. However, in the area of semiconductor sensors, this integration has been limited to Hall effect and optoelectronic devices. The recent combination of micromechanical structures, sensing elements, and signal conditioning is the beginning of a new chapter in sensor technology. The combination of microelectronics with micromechanical structures promises to change future control systems and enable entirely new applications that were previously too costly for commercial purposes.

A sensor with its own dedicated interface circuitry has several advantages. The sensor designer can trade off unnecessary performance characteristics for those that will provide desirable performance advantages to the sensor-interface combination. Normally, interface ICs are designed for a broad range of applications and these tradeoffs are not possible. The combination allows the sensor user to treat the sensor as a "black box" and more easily design a more complex control system.

The integrated sensor takes advantage of integrated temperature sensing to more closely track the temperature of the sensing element and compensate for the effects of temperature over the temperature range. By reducing the number of internal connections, the reliability of integrated sensors is inherently better than a separate sensor and control circuit, even when the separate components are manufactured using a thin-film ceramic substrate. For a 4-terminal sensor element, a reduction from 23 to only 9 connections is possible using an integrated solution. Since the sensor provides the first information provided to a control system, the reliability of this information is critical to the entire system's reliability.

Figure 1.8 is useful in analyzing the results of integrating the first two stages

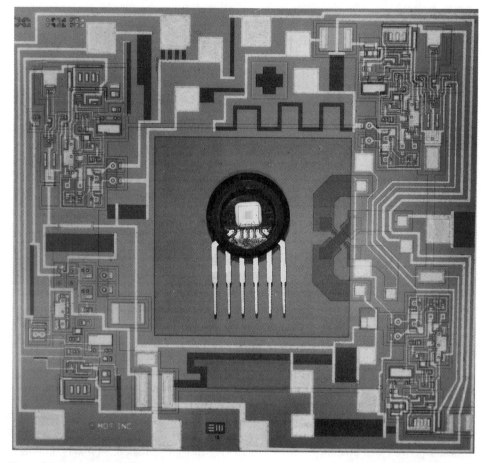

Figure 1.8 Pressure sensor die photomicrograph with packaged sensor over the micromachined diaphragm (*courtesy of* Motorola, Inc.).

mentioned in Figure 1.7 to obtain the sensor portion of a Level III system. The figure shows an enlarged view of a fully integrated piezoresistive pressure sensor with the amplification and signal-conditioning circuitry, including laser-trimmed resistors, located around the silicon diaphragm etched in the middle of the chip. The single piezoresistive element is one arm of the ''X'' that is near the edge of the right-hand side of the diaphragm. The package used for this sensor (shown in the middle of the diaphragm with the actual chip inside) has six terminals; however, only three are required for the application. The other three terminals are used to access measurement points on the chip during the trim process and are not necessary for the sensor to function properly.

One of the major advantages of integrated sensing and signal conditioning is the addition of calibration through laser trimming of thin-film resistors on the sensor die, and the subsequent ability to obtain part-for-part replaceability at the component level. The amount of signal conditioning can vary. For example, the addition of thin-film resistors and laser trimming to the sensing element are all that are necessary to produce a calibrated and temperature-compensated sensor for disposable blood pressure applications. For lower volume applications, the requirement for unique transfer function or interface circuitry may not be cost-effective. In these cases, the basic sensor with external circuitry is still the best choice. Eventually, these unique lower volume applications could also benefit from the advantages of integration.

Sensing and integrated sensing (sensing plus signal conditioning) are somewhat analogous to other mixed-signal processes that exist in semiconductors, especially in power and smart-power technology (i.e., the output side of the control system). Today's smart-power technologies integrate bipolar and CMOS circuitry with multiple power metal oxide semiconductor field effect transistor (MOSFET) output devices. The process is more complex than a discrete power MOSFET, but the performance provided by the combination of technologies to provide a specific function, component reduction for increased reliability, and space reduction for lower cost assemblies more than justifies the higher processing cost. After a number of years of both process and design improvements, smart-power devices have established broad market acceptance, especially for custom designs. Similarly, smarter sensors that provide their own signal conditioning circuitry onboard are in the early phases of market acceptance and are approximately at the same stage of development as smart power was five to seven years ago. However, smart sensors may not have to be custom devices to satisfy a large number of applications.

New packages must be developed for smart sensors to accommodate the additional connections for power, ground, output, diagnostics, and other features that the combination of technologies can provide. Sensors have very few, if any, commonly accepted packages. Instead, each supplier provides a unique pinout and form factor. This problem is exacerbated by the addition of more features to sensors, either through integration or by the addition of circuitry.

Prior to the era of sensor integration, products that combine technology at the package level, rather than the silicon level, have been the industry norm. A hybrid (or module) solution has the advantage of using proven, available technology to achieve a more sophisticated product solution. This may be in a printed circuit board or ceramic substrate form. The steps that can be taken toward a higher level of monolithic integration are shown in Figure 1.9.

A Level IV, two-chip smart sensor is shown in Figure 1.10. This unit uses the integrated pressure sensor showed in Figure 1.8, along with a single chip 8-bit microcontroller unit (MCU) with an onboard A/D converter, and an E^2PROM (electrically erasable, programmable read only memory) to achieve a minimum-component-count smart sensor [9]. Except for three resistors that are used for increased resolution, the other ten components in the circuit are necessary for proper functioning of the MCU. An undervoltage

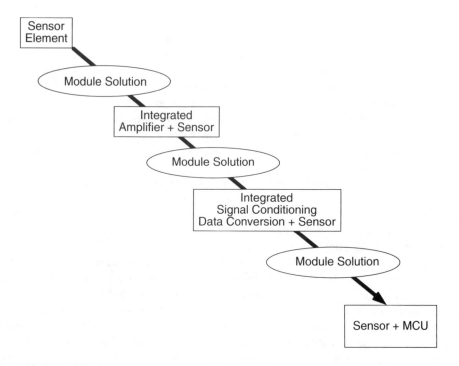

Figure 1.9 Steps of development toward increased integration.

sensing circuit is used for the reset function to provide an orderly powerdown in the case of low battery voltage.

Two jumpers (and *no* potentiometers) are used for the initial calibration. The "01" code for the jumpers (J1 and J2) is used for zero calibration, and a "10" code is used for indicating full-scale value. A "000" on the display indicates zero, and "FFF" indicates the full-scale value during calibration. Either the "00" or "11" code is used for normal operation. The values of zero and full scale are stored in the E^2PROM. Three resistors are used to provide a ratiometric reference of $V_{RL} = 0.3$ to $V_{RH} = 4.7$V utilizing the highest output capability of the integrated sensor.

The liquid crystal display (LCD) is driven directly from the MCU by using the MCU to generate a square wave signal every 32 milliseconds. The backplane is alternately inverted with the segments and out of phase with the segment that is "on." Each of the MCU ports is connected to a digit, and two additional I/O lines are used to generate the decimal point and the backplane signal. This approach utilizes software to eliminate the need for an additional display driver and achieves a reduction in both component count and space. The memory required by the MCU is minimal, only 1K ROM (read only memory) and 4 bytes of E^2PROM, leaving essentially all of the

Figure 1.10 Two-chip smart sensor for pressure measurements (*courtesy of* Motorola, Inc.).

remaining memory for other functions. The program for the MCU is written in C, a high-level language.

1.5 SUMMARY

Today, it is possible to build a smart sensor for several measurements with basically two semiconductor components—the sensor and the MCU. This is the current phase of development and one of the necessary steps to the next level. The "smart money" is being placed on areas that take advantage of the number of technologies that are available or in developments that will fundamentally change the nature of sensing, control systems, and aspects of everyday life. The remaining chapters discuss a variety of aspects of sensing and state-of-the-art developments that will allow those who understand how to apply these developments to create the next generation products and systems.

REFERENCES

[1] Beckwith, T. G., N. L. Buck, and R. D. Marangoni, *Mechanical Measurements*, Reading, MA: Addison-Wesley Publishing Company, 1982.

[2] "Microfabricated Pressure and Acceleration Sensor Terminology," *IEEE Draft Standard, TC-9 Task Group*, 1994.

[3] Wright, C., "Information Conversion Separates Noise Levels So You Can Control Them," *Personal Engineering & Instrumentation News*, April 1993, pp. 63–67.

[4] *Pressure Sensor Data Book*, Motorola Semiconductor Products Sector.

[5] Najafi, K., "Smart Sensors," *Journal of Micromechanics and Microengineering*, Vol. 1, 1991, pp. 86–102.

[6] Ina, O., "Recent Intelligent Sensor Technology in Japan," *Society of Automotive Engineers paper SAE891709*, 1989.

[7] Maitan, J., "Overview of the Emerging Control and Communication Algorithms Suitable for Embedding Into Smart Sensors," *Proc. of Sensors Expo*, Cleveland, OH, Sept. 20–22, 1994, pp. 485–500.

[8] Benson, M. et al., "Advanced Semiconductor Technologies for Integrated Smart Sensors," *Proc. of Sensors Expo*, Philadelphia, PA, Oct. 26–28, 1993, pp. 133–143.

[9] Frank, R., "Two-Chip Approach to Smart Sensors," *Proc. of Sensors Expo*, Chicago, IL, 1990, pp. 104C1–104C8.

Chapter 2
Micromachining

"I wonder how many angels can sit on the head of a pin?"
—Unknown medieval philosopher

2.1 INTRODUCTION

No more than 25,200, given that the pin had a diameter of 1/16 inch (15,875 μm) and each (micromachined) angel had a diameter of 100 μm. So much for philosophy. However, micromachining is causing the reinvestigation of every aspect of physics, chemistry, biology, and engineering. Thermodynamics, mechanics, optics, fluidics, physics, acoustics, magnetics, electromagnetics, and nuclear forces, as well as chemistry and biology, are being investigated in academic, national, and/or industrial research and development (R&D) labs. Micromachining technology is enabling the extension of semiconductor-based sensing beyond temperature, magnetic, and optical effects to produce mechanical structures in silicon and sense mechanical phenomena.

Micromachining is, in the most common usage, a chemical etching process for manufacturing three-dimensional microstructures that is consistent with semiconductor processing techniques. IC manufacturing processes used to make these microstructures include photolithography, thin-film deposition, and chemical and plasma etching. Bulk micromachining has been used to manufacture semiconductor pressure sensors since the late 1970s. Recently, newer techniques such as surface micromachining have been developed that achieve even smaller structures. In addition, the processing techniques for surface micromachining are more compatible with the CMOS processes used to manufacture integrated circuits.

Silicon has many mechanical properties that make it ideal for mechanical structures. As indicated in Table 2.1, it has a modulus of elasticity (Young's modulus) comparable to steel and a higher yield strength than steel or aluminum [1, 2]. Silicon has essentially perfect elasticity, resulting in minimal mechanical hysteresis. Also, silicon's electrical properties have made it the material of choice in most integrated circuits providing established manufacturing techniques for many aspects of micromachined sensors. Mi-

Table 2.1 Properties of Silicon Compared to Other Materials

Property	Silicon	SiC	Diamond	Stainless Steel	Aluminum
Melting Point	1350–1415°C	2830*	3550	1400	660
Max. Operating Temperature	150–175°C (300°C)	873	1100	n/a	n/a
Thermal Expansion	$2.5 \times 10^{-6}/°C$	3.3	1	17.3	25
Thermal Conductivity	1.57 W/cm °C	4.9	20	0.329	2.36
Density	2.3 g/cm^3	3.2	3.5	7.9	2.7
Young's Modulus	1.9×10^{12} dyn/cm^2	7	10.35	2	0.7
Yield Strength	6.9×10^{10} dyn/cm^2	21	53	2.1	0.17
Knoop Hardness	850 kg/m^2	2480	7000	660	130
Dielectric Strength	0.5 MV/cm	4.0	10	n/a	n/a
Bandgap	1.12–1.21 (eV)	3.0	5.47	n/a	n/a

* Sublimation temperature.

After: [1].

cromachined semiconductor sensors take advantage of both the mechanical and electrical properties of silicon. However, products that fully exploit the combination of the mechanical and electrical properties are still in their infancy.

The relative ease of accomplishing both bulk and surface micromachining has led to many researchers investigating a variety of applications. Some of the areas being investigated will lead to smarter sensors through higher levels of integration. The processes used for micromachining and the application of this technology to sensors are the items that will be covered in this chapter.

2.2 BULK MICROMACHINING

Bulk micromachining is a process for making three-dimensional microstructures in which a masked silicon wafer is etched in an orientation-dependent etching solution [3]. Using micromachining technology, several wafers can be fabricated simultaneously and lot-to-lot consistency is maintained by controlling a minimal number of parameters. Key parameters in bulk micromachining include crystallographic orientation, etchant, etchant concentration, semiconductor starting material, temperature, and time. Photolithography techniques common in IC technology precisely define patterns for etching both sides of silicon wafers. The crystallographic orientation, etchant, and semiconductor starting material are chosen by design, leaving etchant concentration, temperature, and time as lot-to-lot control items.

Cubic silicon has a total of 26 faces: 18 faces are square and 8 are equilateral triangles (indicated by the <111> plane). Silicon ICs are typically fabricated (manufactured) using <100> or <111> silicon. In bulk micromachining, an anisotropic (unidirectional) etchant, such as ethylene-diamine pyrocatechol (EDP), hydrazine, or potassium hydroxide (KOH) attacks the <100> plane of silicon. The <100> plane is etched at a much faster rate than the <111> plane, typically 100 times faster. N-type silicon is etched at a much faster rate (>50 times faster) than P-type, so N-type material is often used as the starting material. P-type material can be epitaxially grown on the wafer or diffused into the wafer to add a further control element (an etch stop) in defining the dimensions. Agitation maintains uniform concentration during anisotropic etching. The characteristic shape (preferential etching) of anisotropic etching of <100> silicon is shown in the cross-section of Figure 2.1(a), which results in a 54.7-deg angle for the <100> silicon [4]. The top view of etching into the surface of the silicon appears as a pyramid-shaped pit.

Etch rates of 1.0–1.5 μm/min occur in the <100> plane of silicon with etch temperatures of 85°C to 115°C for common etchants [5]. Isotropic etching, shown in Figure 2.1(b), has etch rates independent of the crystallographic orientation. Isotropic etching allows undercutting and cantilever structures to be produced. However, in bulk silicon it is more difficult to control than anisotropic etching. Undercutting and suspended structures are achieved with anisotropic etching with mask patterns and extended etch times. To expand the applications of micromachining, other chemicals, such as sodium hydroxide (NaOH) and hydrofluoric acid (HF), are used for etching the <100> and <110> plane. Also, other materials such as polysilicon can be used. Table 2.2 lists the etch rates of common materials and possible etchants [6].

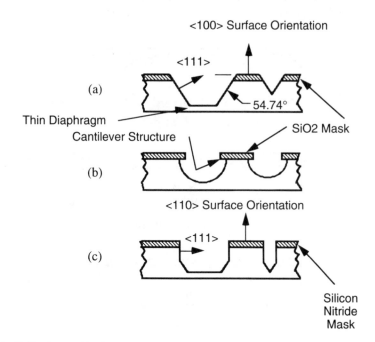

Figure 2.1   Bulk-micromachined structures: (a) anisotropic, (b) isotropic, and (c) alternative crystal orientation and mask material.

Table 2.2 Attributes of Etchants

Etched Material	Etch Rate (nm/min.)	Etchant	Comments
Al	80	$SiCl_4$	—
Polysilicon	60	$SiCl_4$	—
Si_3N_4	50	$SiCl_4$	—
SiO_2	25	$SiCl_4$	—
SnO_2	20	$SiCl_4$	—
ZnO	5	$SiCl_4$	—
<100>silicon	1250	EDP	Anisotropic
<110>silicon	1400	KOH	Anisotropic
Si	900–1300	SF_6	Isotropic
Si	10000–300000	HF	Isotropic
SiO2	400-8600	HF	Surface

After: [2].

Etch stop techniques enhance the accuracy of wet chemical etching. The most common techniques for etch depth control in bulk micromachining are shown in Figure 2.2(a-c) [5]. Precisely controlled diffusions (a), epitaxially grown layers in the silicon crystal (b), or field-enhanced depletion layers (c) inhibit the etching process, allowing very

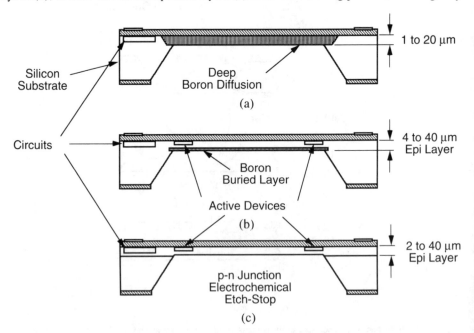

Figure 2.2 Etch stop techniques: (a) deep boron diffusion, (b) boron buried layer, and (c) p-n junction.

accurate structures to be obtained. Boron etch stops can be used to produce layers as thick as 15–50 μm with oxide masking capable of protecting other areas of the chip for adding circuitry.

In addition to micromachining, different types of wafer bonding are needed to produce more complex sensing structures. The attachment of silicon to a second silicon wafer or silicon to glass is an important aspect of semiconductor sensors. Two pressure sensor examples explain the difference in these bonding techniques.

2.2.1 Silicon on Silicon Bonding

A frequently used approach for manufacturing semiconductor pressure sensors uses a bulk-micromachined diaphragm anisotropically etched into a silicon wafer. Piezoresistive sensing elements diffused or ion-implanted into the thin diaphragm are either a four-element Wheatstone bridge, or a single element positioned to maximize the sensitivity to shear stress. Two silicon wafers are often used to produce the piezoresistive silicon pressure sensor.

Figure 2.3 shows a two-layer silicon on silicon pressure sensor [7]. The top wafer is etched until a thin square diaphragm approximately one mil (25.6 μm) in thickness is achieved. The square area and the 54.7-deg angle of the cavity wall are extremely reproducible. In addition to a sealed reference cavity for absolute pressure measurements, the two-layer silicon sensor allows atmospheric or a reference pressure to be applied to one side of the sensor by an inlet hole, which is micromachined in the silicon bottom (constraint) wafer. Several methods are used to attach the top wafer to the bottom, including anodic bonding, glass frit seal, and direct wafer (silicon to silicon) bonding, or silicon fusion bonding.

The sensor shown in Figure 2.3 uses a glass frit to attach the top wafer to the bottom wafer. A glass paste is applied to the bottom (constraint) wafer, which is then thermo-

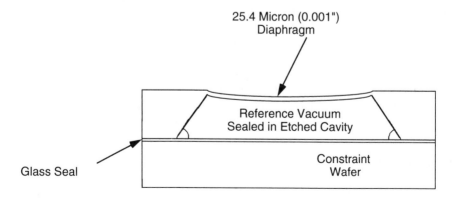

Figure 2.3 Pressure sensor with silicon to silicon bonding.

compression bonded to the top wafer containing the bulk-micromachined pressure sensing structure. The bottom wafer, containing the glass, provides stress isolation and allows a reference vacuum to be sealed inside the combined structure.

To extend the pressure capability to very low pressure readings ≤ 2 inches or 5 cm of water) and minimize nonlinearity, several different approaches are being pursued, including: silicon bosses used as stress concentrators in circular, square, and dual rectangular diaphragms; convoluted square diaphragm; and etch stop techniques to control the diaphragm thickness. Diaphragms as thin as 2.5 μm have been used to produce capacitive pressure sensors for 300 mtorr and lower pressure applications [8]. Micromachining can be enhanced by using the electronics capability inherent from semiconductor manufacturing. This may provide an additional solution for low-pressure measurements.

2.2.2 Silicon on Glass (Anodic) Bonding

Electrostatic or anodic bonding is a process used to attach a silicon top wafer to a glass substrate, and is also used to attach silicon to silicon. Anodic bonding attaches a silicon wafer, either with or without an oxidized layer, to a borosilicate (Pyrex®) glass heated to about 400°C when 500V or more are applied across the structure [4]. An example of a product manufactured with anodically bonded silicon to glass is the silicon capacitive absolute pressure (SCAP) sensing element shown in Figure 2.4 [9].

The micromachined silicon diaphragm with controlled cavity depth is anodically bonded to a Pyrex® glass substrate, which is much thicker than the silicon die. Feed-through holes are drilled in the glass to provide a precise connection to the capacitor plates inside the unit. The glass substrate is metalized using thin-film deposition techniques and photolithography defines the electrode configuration. After attaching the top silicon wafer to the glass substrate, the drilled holes are solder-sealed under vacuum. This provides a capacitive sensing element with an internal vacuum reference and solder bumps for direct mounting to a circuit board or ceramic substrate. The value of the capacitor changes linearly from approximately 32 to 39 pF with applied pressure from 17 to 105 kPa. The capacitive element is 6.7 mm by 6.7 mm and has a low temperature coefficient of capacitance (-30 to 80 ppm/°C), good linearity (≈ 1.4%), fast response time (≈ 1 ms), and no exposed bond wires.

Silicon is also bonded to a second silicon wafer using anodic bonding. A thin (≈ 4 μm) layer of Pyrex® glass is sputtered on one of the layers and a much lower voltage of approximately 50V is applied [4]. Silicon instead of glass for the support structure allows additional wet-etching techniques to be performed later in the process.

2.3 SURFACE MICROMACHINING

The selective etching of multiple layers of deposited thin films, or surface micromachining, allows movable microstructures to be fabricated on silicon wafers [10]. With surface

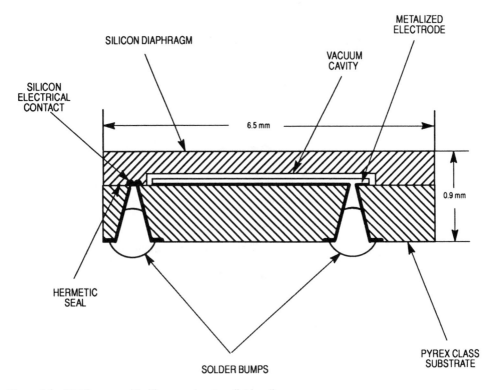

Figure 2.4 SCAP sensor with silicon on glass (anodic) bonding.

micromachining (shown in Figure 2.5), layers of structural material (typically polysilicon and a sacrificial material such as silicon dioxide) are deposited and patterned. The sacrificial material acts as an intermediate spacer layer and is etched away to produce a free-standing structure. Surface-micromachining technology allows smaller and more complex structures with multiple layers to be fabricated on a substrate. However, annealing is required to reduce stresses in the layers, which can cause warping. In contrast, bulk micromachining is typically stress free.

Surface micromachining has been used to manufacture an accelerometer for automotive air bag applications. A three-layer differential capacitor is created by alternate layers of polysilicon and phosphosilicate glass (PSG) on a 0.015-in (0.38 mm) thick 4-in (100 mm) wafer [11]. A silicon wafer serves as the substrate for the mechanical structure. The trampoline-shaped middle layer is suspended by four support arms. This movable structure is the seismic mass for the accelerometer. The upper and lower polysilicon layers are fixed plates for the differential capacitor. The glass is sacrificially etched by an isotropic etch, such as hydrofluoric acid (HF).

Because of the small spacing (≈ 2 μm) that is possible with surface micromachining,

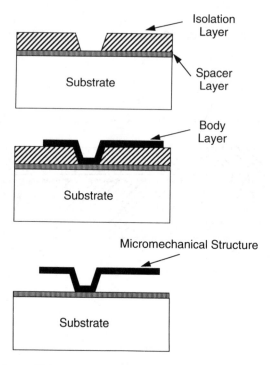

Figure 2.5 Surface micromachining steps.

new issues arise that affect both the sensor design and the manufacturing process. Squeeze-film damping, stiction, and particle control must be addressed in each new design. The next three sections will describe these areas and some approaches being used to deal with them.

2.3.1 Squeeze-Film Damping

The movement of structures separated by only a few microns can be greatly affected by the actual spacing and ambient (gas or vacuum) between the structures. This effect is known as squeeze-film damping. Squeeze-film damping can be significant in bulk-micromachined capacitive structures where closer spacing is needed to achieve higher capacitance values. It is inherent in surface micromachining where spacing is only a few microns. For a particular structure, the gas that separates the layers has a viscous damping constant that increases with the inverse cube of the spacing [12]. Incorporating holes in the surface-micromachined structure allows the damping to be tuned for desired characteristics. Holes also provide distributed access for the etchant to reduce etching time and the possibility of overetching portions of the structure.

2.3.2 Stiction

Static friction, or stiction, is a phenomenon that occurs in surface micromachining resulting from capillary forces generated during the wet-etching of the sacrificial layers [13]. Under certain fabrication conditions, these microstructures can collapse and permanently adhere to the underlying substrate. The failure is catastrophic and must be prevented to achieve high process yield and a reliable design. Preventing the top structure from contacting the bottom structure requires minimizing the forces acting on the device when the liquid is withdrawn or minimizing the attractive forces between the structures if they contact each other. Techniques used to prevent stiction depend on the manufacturer, the product design, and the process flow [10].

2.3.3 Particulate Control

One of the design problems that must be solved in working with structures separated by only a few microns is avoiding contamination. Wafer-level packaging is an attractive solution because it provides a low cost, protective, and safe environment for moving parts that require additional electrical testing and assembly processes. The package protects the device from microscopic particulates, handling, and provides an ambient atmosphere for adjusting damping. A hermetically sealed accelerometer chip that can be overmolded in a low-cost conventional epoxy package provides an example of one approach to avoiding contamination.

The general concept of a sensing die with moving parts that needs protection from the environment is shown in Figure 2.6 [11]. The three polysilicon layers are sealed inside

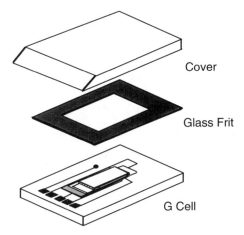

Cover

Glass Frit

G Cell

Figure 2.6 Combined surface and bulk micromachining in an accelerometer.

a protective cavity formed by the silicon substrate, a bulk-micromachined top (cover) wafer, and a glass layer that entirely surrounds the polysilicon structure. The glass is spaced a distance from the polysilicon structure to avoid the possibility of mechanical interference. The glass serves not only as the bonding media, but also as the "mechanical" spacer that provides the "elbow room" for the movable structure. A 0.015-in silicon wafer is used as the top or cover wafer. The glass is applied to the top wafer, which is then thermocompression bonded to the bottom wafer containing the micromachined accelerometer structures. The top wafer design provides a hermetic environment, physical protection, and access to the bond pads. When bonded, a sealed cavity for controlled squeeze film damping is achieved.

Other techniques to minimize particle contamination use metal can or ceramic packaging. In these cases, the final package provides a hermetic environment for the structure. Prior to packaging, attention must be given to other processing steps that could allow particles to be trapped in the structure.

2.3.4 Combinations of Surface and Bulk Micromachining

The previous example combines surface and bulk micromachining. Another combination of the two etching techniques has been reported. An airgap capacitive pressure sensor has been demonstrated that combines bulk and surface micromachining on a single wafer [14]. The structure in Figure 2.7 used standard IC processing to create NMOS circuits with an additional polysilicon layer to produce a capacitor with a 0.6-μm thick dielectric. Surface micromachining allowed a smaller gap to be produced. The MOS circuitry on the top of the wafer was not exposed to the pressure media. The inlet for the pressure source and the release for the surface micromachined structure were bulk micromachined into the silicon substrate using a KOH etch. A sensitivity of 0.93 mV/kPa (6.4 mV/psi) was measured for the 100 by 100-μm capacitive structure. The 100-fF (femtofarad = 10^{-15}) capacitor had a resolution of 30 attofarads (10^{-18}).

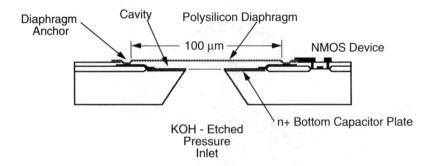

Figure 2.7 Combined surface and bulk micromachining in a pressure sensor.

2.4 OTHER MICROMACHINING TECHNIQUES

The micromachining techniques discussed so far have been wet-etching techniques used primarily for piezoresistive and capacitive sensing for measuring pressure and acceleration. These are the basic processes used in research that have progressed into commercial manufacturing. However, other techniques are being developed that overcome the limitations, extend the types of measurements that can be made, or enhance the capabilities of previously discussed approaches. These newer techniques include the LIGA process, several methods of dry etching, silicon fusion bonding, and laser micromachining.

2.4.1 LIGA Process

One of the newer micromachining processes is the LIGA (derived from German terms for lithography, electroforming, and molding) process which combines X-ray lithography, electroforming, and micromolding techniques [15]. The LIGA process allows high-aspect-ratio (height/width) structures to be fabricated. The X-ray patterned photoresist molds are chemically etched in a metal plate. A polyimide layer a few microns thick acts as a sacrificial layer. A complementary structure is built up by electrodepositing a metal layer, such as nickel. After the final etching process, portions of the microstructure remain attached to the substrate and are able to move freely. Temperatures are under 200°C for the entire process. Tiny 100-μm gears (about the diameter of a human hair) have been made by this process. Up to a dozen metal gears have been driven by a low-level magnetic field powered micromotor. The LIGA process greatly expands micromachining capabilities, making possible vertical cantilevers, coils, microoptical devices, microconnectors, and actuators [10].

A differential pressure sensor with double-sided overload protection has been fabricated with a modified LIGA process [16]. As shown in Figure 2.8, this design combines isotropic bulk micromachining for the cavity and flow channels, LIGA processing for the electroplated nickel structure (which is 100 μm thick and has a gap of 0.80 μm), and sacrificial etching for the polysilicon diaphragm. The high-aspect-ratio metallic stop limits motion and suppresses diaphragm stress, as well as facilitating the option of a second signal to verify performance.

2.4.2 Dry Etching Processes

Plasma etching and reactive ion etching can produce structures that are not possible from the wet chemical etching processes. Plasma etching is an etching process that uses an etching gas instead of a liquid to chemically etch a structure. Plasma-assisted dry etching is a critical technology for manufacturing ultralarge-scale integrated circuits [17]. Plasma etching processes are divided into four classes: sputtering, chemical etching, energetic or

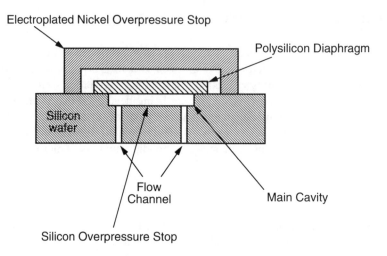

Figure 2.8 LIGA process with sacrificial etching for differential pressure sensor.

damage-driven ion etching, and sidewall inhibitor ion-assisted anisotropic etching. Figure 2.9 illustrates the different mechanical structures that result from these techniques.

Manufacturing control of a plasma etching process addresses etch rate control, selectivity control, critical dimension control, profile control, and control of surface quality and uniformity. The control of these parameters allows new structural elements to be achieved. Plasma etching was used in the surface-micromachined sensor in Figure 2.7. SF_6 was used to etch LPCVD (low-pressure chemical vapor deposition) silicon nitride, and polysilicon (CF_4) was used to etch BPSG (boro-phospho-silicate glass), and CH_4 was used to etch BPSG and the nitride. Plasma etching was also performed on the aluminum metal [14].

Ion beam milling is a dry etching process that uses an ion beam to remove material through a sputtering action. It can be used separately or with plasma etching. When it is combined with plasma etching, it is also known as reactive ion etching (RIE).

A reactive ion etching process has been used for mechanical structures and capacitors in single-crystal silicon [18]. The RIE process was used to achieve high-aspect-ratio mechanical structures difficult to achieve using surface micromachining and wet-etch bulk micromachining. Structures with feature sizes down to 250 nm and with arbitrary structure orientations on a silicon wafer have been produced using the single crystal reactive etching and metalization (SCREAM) process. As shown in Figure 2.10, a thick-film deposition is not required. A silicon cantilever beam is formed with aluminum electrodes on both sides of the beam using silicon dioxide for insulation and for the top and sidewall etching mask. The SCREAM process can be used to fabricate complex circular, triangular, and rectangular structures in single-crystal silicon.

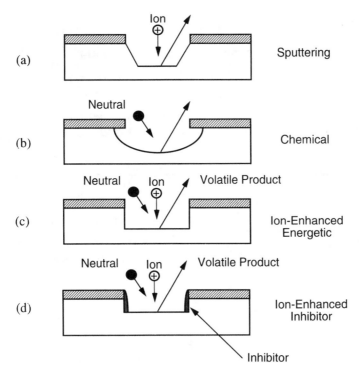

Figure 2.9 Different structures from plasma etching: (a) sputtering, (b) chemical, (c) ion-enhanced energetic, and (d) ion-enhanced inhibitor.

2.4.3 Silicon Fusion Bonding

A technique that bonds wafers at the atomic level without polymer adhesives or an electric field is known as silicon fusion bonding (SFB) or direct wafer bonding (DWB). Before bonding, both wafers are treated in a solution, such as boiling nitric acid or sulfuric peroxide [19]. This step covers the surface of both wafers with a few monolayers of reactive hydroxyl molecules. Initial contact of the wafers holds them together through strong surface tension. Subsequent processing at temperatures from 900°C to 1,100°C drives off the hydroxyl molecules. The remaining oxygen reacts with the silicon to form silicon dioxide and fuses the two surfaces together.

Silicon fusion bonding can be used to reduce the size of a micromachined structure. As shown in Figure 2.11, the anisotropically etched cavity can be much smaller and yet the diaphragm area is identical for an SFB pressure sensor compared to a conventional pressure sensor. The bottom wafer with the anisotropically etched cavity is silicon fusion bonded to a top wafer. After bonding, the top wafer is etched back to form a thin diaphragm and the bottom wafer is ground and then polished to open the access to the diaphragm. For

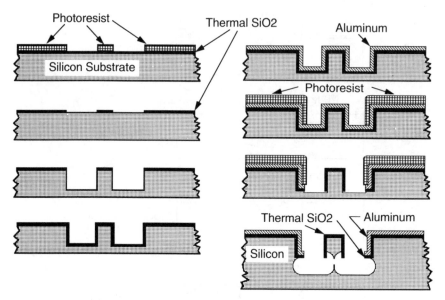

Figure 2.10   Reactive ion etching in the SCREAM process.

absolute-pressure sensors, such as the units in Figure 2.11, a support wafer is attached to the structure containing the diaphragm produced by any of the previously described methods.

2.4.4 Lasers in Micromachining

In addition to chemical etching, lasers are used to perform critical trimming and thin-film cutting in semiconductor and sensor processing. The flexibility of laser programming systems allows their usage in marking, thin-film removal, milling, and hole drilling [20]. Lasers also provide noncontact residue-free machining in semiconductor products including sensors. The precise value of the thin-film resistors in interface circuits is accomplished by interactive laser trimming. Interactive laser trimming at the die level for micromachined sensors has been used to manufacture high volume, interchangeable, calibrated, and compensated pressure sensors since the mid 1980s [21].

Lasers have been used to drill through silicon wafers as thick as 0.070 inch (1.78 mm) with hole diameters as small as 0.002 inch (50.8 μm) [20]. For example, 5-mil (127 μm) holes spaced on 10-mil (254 μm) centers have been drilled into a 15-mil (0.381 mm) thick wafer. The hole diameters and close spacing are achieved without causing fracturing or material degradation. Also, lasers can vaporize the material (ablation) using high-power density.

Lasers have also been investigated as a means of extending the bulk-micromachining

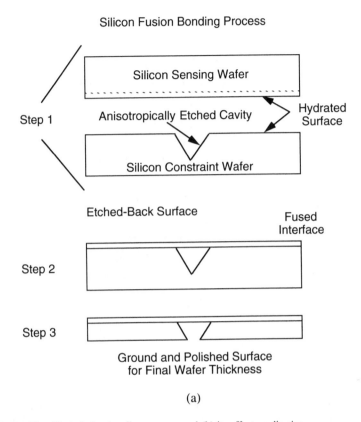

Figure 2.11 (a) The silicon fusion bonding process, and (b) its effect on die size.

process [22]. Figure 2.12 shows that either <110> or <100> wafers can be processed using a combination of photolithography, laser melting, and anisotropic etching. A deeper and wider etch occurs in the area that has been damaged. The grooved shape or microchannel obtained by this process has been used to precisely position fibers and spheric lenses in hybrid microoptical devices without requiring additional bonding or capturing techniques.

2.4.5 Post Etching to Obtain Smarter Structures

One of the main concerns in integrating circuitry with micromachined structures is the effect that the etching and the resulting temperatures can have on other semiconductor processes and the circuit elements. The combined surface and bulk-micromachining process shown Figure 2.7 also included NMOS circuitry to process the signal for the capacitive sensor. Other examples will be discussed to show the different approaches that are being pursued.

The process for an integrated sensor is more complex since it involves the joining

Conventional Absolute Pressure Sensor

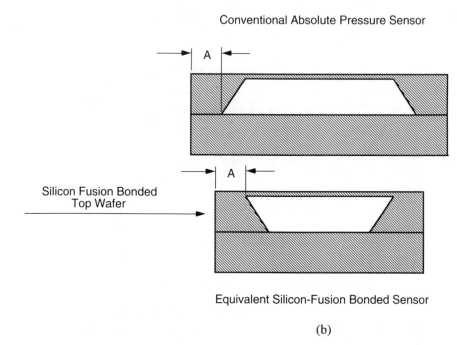

Silicon Fusion Bonded
Top Wafer

Equivalent Silicon-Fusion Bonded Sensor

(b)

Figure 2.11   Continued.

of unique sensor processes with traditional IC processes. For example, bipolar operational amplifiers (op amps) are typically fabricated on <111> silicon substrates while pressure sensors are typically fabricated on <100> silicon substrates. A production solution to this problem uses a standard op amp design on <100> silicon substrates [23]. While this reduces the breakdown voltage in the op amp, it does not result in breakdown voltages less than the 10V required for the sensor. The <100> silicon substrate is P-type with an N-type epitaxial layer. This N-type epitaxial layer matches both the op amp and the pressure sensor requirement. The pressure sensor also has several unique processes, such as cavity etch and a thin-film requirement that must be combined with the op amp processing.

The task of combining two different process flows is complicated by allowances that must be made for each process. The final combination allows the diffusion of the sensor to remain the same and preserves the general process flow of the op amps. The cavity etch process defines and etches the silicon to produce a diaphragm in the silicon. The etch process calls for several masking steps to protect the silicon during the etch. These layers must be added into the flow in a manner that allows a proper layer structure.

The manufacturing steps for the integrated sensor include a thin-film process that serves as the circuit link between the pressure sensor and the op amp. The thin film is patterned to make laser-trimmable resistors. These relatively high resistance, but low

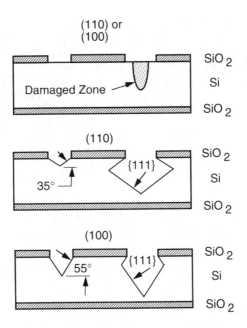

Figure 2.12 Combined effect of laser micromachining and anisotropic etching.

temperature coefficient of resistance (TCR) metal films have allowed resistor networks to be manufactured economically on chip. Each unit is laser-trimmed individually to the proper offset, sensitivity, and temperature characteristics. The use of these techniques enable robust, cost-effective, and precise integrated sensors to be manufactured in high volume. Figure 2.13 shows a scanning electron microscope (SEM) view of the single piezoresistive sensing element and circuitry that was integrated in the final process.

Standard commercial CMOS foundries, through the MOSIS service, provide the capability to have CMOS circuitry and postprocessing EDP maskless etching to yield integrated circuitry and mechanical structures on the same silicon wafer [24]. The micro-machining is performed after the CMOS processing has been completed. This allows mixed electrical and mechanical silicon devices to be developed and even produced in small volumes. With this approach, instrumentation and measurement devices can be reduced in both size and cost. High-volume cost-sensitive applications, such as automobiles or ap-pliances, will require a dedicated wafer fabrication facility under the direct control of the sensor manufacturer.

Another standard 3-μm CMOS process has been combined with anisotropic wet-etching at the Fraunhofer Institute in Berlin [25]. The parameters of the CMOS process were not modified. An additional six masking layers were needed to fabricate monolithic accelerometers. The sensor structure was defined in the CMOS processing and testing of the IC portion was possible prior to the micromachining steps. Both top and backside (with etch stop) anisotropic etching were performed to create the accelerometer structure.

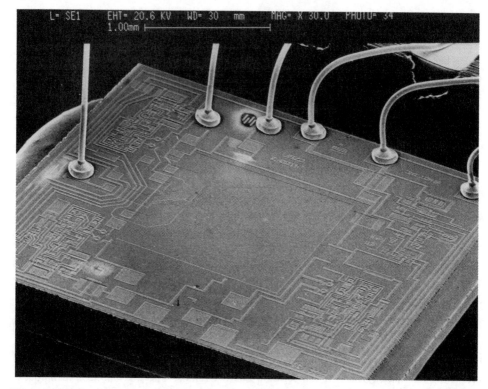

Figure 2.13 Monolithic piezoresistive pressure sensor (*courtesy of* Motorola, Inc.).

A modular CMOS and surface micromachining process was developed at the University of Berkeley in California [26]. The highly optimized CMOS process can be performed by a silicon foundry. The aluminum interconnects in the CMOS process were replaced with tungsten and a TiN/TiSi$_2$ diffusion layer at metal-silicon contacts was added to increase the temperature ceiling for the process. Oxide and nitride films were deposited to isolate the CMOS from the polysilicon.

For monolithic integrated sensors, processing temperatures are a major concern. With bulk micromachining, the same processing temperature constraints apply to both the sensor and control electronics. Using surface micromachining techniques, the structures are dielectrically isolated so that leakage is not a problem at elevated temperatures. The maximum operating temperature is therefore limited by the control electronics.

Approaches for integration at the component level can be totally monolithic or multichip modules with separate dice for the sensor and the smart interface. Microelectronics Center of North Carolina (MCNC) has investigated flip-chip technology (see Chapter 10 for more details) to add circuitry to processed micromachined devices with solder bumping [27]. The approach has been studied for surface micromachining and a combined surface and bulk-micromachining process. Substrate materials including silicon, Pyrex®, quartz, and GaAs are being explored.

Die size and process complexity are basic drivers of cost for semiconductor processes *and* for sensors. The smaller the die size, the greater number of dice that fit on a particular wafer. The simpler the process, the lower the cost for processing and the higher the yield. Combined processes must achieve cost reduction, performance advantage(s), or both to be competitive in the marketplace.

2.5 OTHER MICROMACHINED MATERIALS

The most common IC materials (Si, SiO_2, Al, and Si_3N_4) have played an important role in established micromachining technology. However, improving the performance of a sensor for a particular application, addressing a higher operating range, or sensing a new parameter can require other materials and sensing techniques. Table 2.1, shown earlier in this chapter, provides a comparison of GaAs, SiC, and diamond to silicon. Silicon begins to exhibit plastic deformation at temperatures above 600°C. Higher operating temperatures are among the desirable properties that make these alternate materials attractive for ICs *and* semiconductor sensors. Other semiconductor materials, such as indium phosphide (InP), are also being investigated for micromachined sensors. Also, several metals, metal oxides, and polymer films are being used in micromachined devices. Three examples show the variety of approaches being pursued.

2.5.1 Diamond as an Alternate Sensor Material

Selectively deposited diamond film has been used as the thermal element in a flow sensor [28]. The bulk-micromachined structure with diamond film is shown in Figure 2.14. Processing for the boron-doped heater resistors is performed at temperatures above 1,000°C. After surface preparation, the diamond film is grown in the desired regions by microwave plasma chemical vapor deposition in a mixture of hydrogen and methane gas and a substrate temperature of 900°C. The aluminum pad contact to the resistors is the next

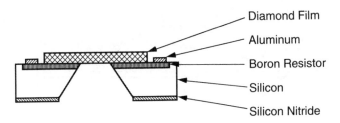

Figure 2.14 Diamond film flow sensor.

step. Finally, the back of the wafer is anisotropically etched to form the bridge structure for the flow sensor. Besides higher operating temperatures, diamond's resistance to corrosive and abrasive environments makes it attractive as a flow sensor.

2.5.2 Metal Oxides and Piezoelectric Sensing

In addition to semiconductor materials, various metal oxides can be deposited on micromachined structures. A pressure sensor with a zinc oxide (ZnO) piezoelectric sensing element has been made in an IC-compatible process [29]. Surface micromachining is used to create the cavity. As shown in Figure 2.15, a thermal oxide layer (Tox) is grown to isolate the sensor from the silicon substrate. The lower polysilicon electrode is encapsulated in Si_3N_4. The spacer oxide layer is sacrificially etched. Polysilicon is used to form an electrically conductive structural support and cavity for the sensor. The 0.95-μm thick active ZnO layer is deposited by RF sputtering. In addition to forming the active piezoelectric film, the sputtering also seals the sidewalls of the structure with an O_2-Ar mixture at 10 mtorr inside the cavity. The sensor exhibits approximately 0.36 mV/μbar sensitivity and 3.4-dB variation over the range 200 Hz to 40 kHz. The dimensions of the sensor range from 50 by 50 μm^2 to 250 by 250 μm^2.

2.5.3 Films on Microstructures

The sensitive layers for chemical sensors can be deposited over a bulk-micromachined structure. As shown in Figure 2.16, one approach to a semiconductor gas sensor uses a thin metal oxide semiconductor deposited over an integrated heating element [30]. The resistivity of the metal oxide depends upon the reducing or oxidizing environment being sensed. The sensitive area is thermally insulated from the silicon substrate to minimize power consumption. Response time of the device is less than 1 sec with an operating temperature range of 250°C to 400°C. The structure is contained in a 3.5-mm by 3.5-mm by 0.3-mm chip.

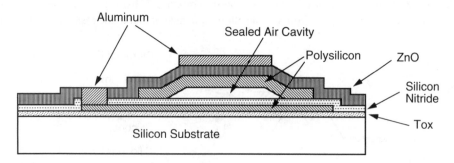

Figure 2.15 Piezoelectric surface micromachined sensor.

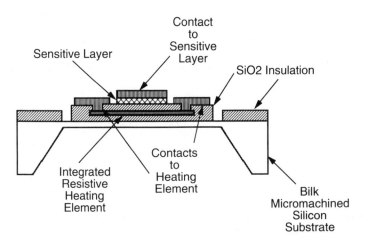

Figure 2.16 Sensing layer deposited over bulk-micromachined structure.

Polymer films are also being deposited on both surface and bulk-micromachined structures. The films act as a membrane in biosensors. Multiple sites can have a variety of films sensitive to different levels of a particular substance or to different substances. These sensors call for a very small sample to provide an analysis, making them ideal for usage on infants and small children or any application where a small sample is available.

2.6 SUMMARY

The trends in the sensing industry for micromachined structures include the following:

· Improved accuracy and resolution;
· Extended operating range of existing sensors;
· The development of higher operating temperature sensors;
· More complex structures using advanced micromachining technology and new materials;
· The addition of integrated circuits;
· Cost reduction through process improvements and die shrinking.

More complex structures will enable new parameters to be sensed and combined measurements to be performed. The continued merging of electronics and various micromachining techniques promises to be a key factor for smart sensing.

REFERENCES

[1] Bryzek, J., K. Peterson, and W. McCulley, "Micromachines on the March," *IEEE Spectrum*, May 1994, pp. 20–31.

[2] Chang, S. S., D. B. Hicks, and R. C. O. Laugal, "Patterning Zinc Oxide Films," *Technical Digest IEEE Solid-State Sensor and Actuator Workshop,* June 22–25, 1992, Hilton Head, SC, pp. 41–45.

[3] Howe, R. T., R. S. Muller, K. J. Gabriel, and W.S.N. Trimmer, "Silicon Micromechanics: Sensors and Actuators on a Chip," *IEEE Spectrum,* July 1990, pp. 29–35.

[4] Goodenough, F., "Sensor ICs: Processing, Materials Open Factory Doors," *Electronic Design,* April 18, 1985, pp. 132–148.

[5] Wise, K. D., "VLSI Circuit Challenges for Integrated Sensor Systems," *IEEE 1990 Symposium on VLSI Circuits,* pp. 19–22.

[6] Chang, S.-C., D. B. Hicks, and R.C.O. Laugal, "Patterning of Zinc Oxide Thin Films," *IEEE 92TH0403-6 from Solid-State Sensor and Actuator Workshop,* Hilton Head, SC, June 22–25, 1992, pp. 41–45.

[7] Frank R., and J. Staller, "The Merging of Micromachining & Microelectronics," *Proc. of the 3rd International Forum on ASIC and Transducer Technology,* Banff, Alberta, Canada, May 20–23, 1990, pp. 53–60.

[8] Zhang, Y., S. B. Crary, and K. D. Wise, "Pressure Sensor and Simulation Using the CAEMEMS-D Module," *IEEE Solid-State Sensors and Actuators Workshop,* Hilton Head, SC, June 4–7, 1990, pp. 32–35.

[9] Behr, M. E., C. F. Bauer, and J. M. Giachino, "Miniature Silicon Capacitance Absolute Pressure Sensor," *Third International Conference on Automotive Electronics,* London, U.K., Oct. 22–23, 1981, pp. 255–260.

[10] Ristic, Lj., *Sensor Technology and Devices,* Norwood, MA: Artech House, 1994.

[11] Ristic, Lj. et al., "A Capacitive Type Accelerometer with Self-Test Feature Based on a Double-Pinned Polysilicon Structure," *Digest of Technical Papers for Transducers '93,* June 1993, pp. 810–813.

[12] Suzuki, S. et al., "Semiconductor Capacitance-Type Accelerometer with PWM Electrostatic Servo Technique," *SAE Sensors and Actuators 1991 P-242,* Warrendale, PA, pp. 51–57.

[13] Mastrangelo, C. H., and C. H. Hsu, "Mechanical Stability and Adhesion of Microstructures Under Capillary Forces-Part I: Basic Theory and Part II: Experiments," *Journal of Microelectromechanical Systems,* March 1993, pp. 33–55.

[14] Kung, J. T., and H. S. Lee, "An Integrated Air-Gap Capacitor Pressure Sensors and Digital Readout with Sub-100 Attofarad Resolution," *Journal of Microelectromechanical Systems,* Sept. 1992, pp. 121–129.

[15] Gosch, J., "Deep-Etch Lithography Yields 0.5 mm Diameter Motor," *Electronic Design,* April 16, 1992, pp. 34–38.

[16] Choi, B. et al., "Development of Pressure Transducers Utilizing Deep X-ray Lithography," *IEEE 91CH2817-5 from Transducers '91,* pp. 393–396.

[17] Flamm, D. L., "Feed Gas Purity and Environmental Concerns in Plasma Etching," *Solid State Technology,* Oct. 1993, pp. 49–54.

[18] Zhang, Z. L., and N. C. McDonald, "A RIE Process for Submicron, Silicon Electromechanical Structures," *Journal of Micromechanics and Microengineering,* March 1992, pp. 31–38.

[19] Peterson, K., and P. Barth, "Silicon Fusion Bonding: Revolutionary Tool for Silicon Sensors and Microstructures," *Proc. of Wescon '89,* pp. 220–224.

[20] Swenson, E. J., "Laser Micromachining for Circuit Production," *Microelectronic Manufacturing and Testing,* March 1990, pp. 17–18.

[21] Cumberledge, W., and J. M. Staller, *IEEE Sensors and Actuators,* Hilton Head, SC, 1986.

[22] Alavi, M. et al., "Laser Machining of Silicon for Fabrication of New Microstructures," *IEEE 91CH2817-5 from Transducers '91,* pp. 512–515.

[23] Baskett, I., R. Frank, and E. Ramsland, "The Design of a Monolithic Signal Conditioned Pressure Sensor," *IEEE 91CH2994-2 CICC '91,* pp. 27.3.1–273.4.

[24] Marshall, J. C., M. Parameaswaran, M. E. Zaghoul, and M. Gaitan, "High-Level CAD Melds Micromachined Devices with Foundries," *IEEE Circuits and Devices,* Nov. 1992, pp. 10–17.

[25] Riethmueller, W., W. Benecke, U. Schnakenberg, and B. Wagner, "Development of Commercial CMOS Process-Based Technologies for the Fabrication of Smart Accelerometers," *IEEE 91CH2817-5 from Transducers '91,* pp. 416–419.

[26] Yun, W., R. T. Howe, and P. R. Gray, "Surface Micromachined, Digitally Force-Balanced Accelerometer

with Integrated CMOS Detection Circuitry," *IEEE Solid-State Sensors and Actuators Workshop*, Hilton Head, SC, June 22–25, 1992, pp. 126–131.

[27] Markus, K. W., V. Dhuler, and A. Cowen, "Smart MEMS: Flip Chip Integration of MEMS and Electronics," *Proc. of Sensors Expo*, Cleveland, OH, Sept. 20–22, 1994, pp. 559–564.

[28] Ellis, C. D. et al., "Polycrystalline Diamond Film Flow Sensor," *IEEE Solid-State Sensors and Actuators Workshop*, Hilton Head, SC, June 4–7, 1990, pp. 132–134.

[29] Schiller, P., D. L. Polla, and M. Ghezzo, "Surface-Micromachined Piezoelectric Pressure Sensors," *IEEE Solid-State Sensors and Actuators Workshop*, Hilton Head, SC, June 4–7, 1990, pp. 188–190.

[30] Microsens Brochure, Neuchatel, Switzerland.

Chapter 3

The Nature of Semiconductor Sensor Output

"What a revolting development this is!"
—Daffy Duck, from Warner Bros. cartoons

3.1 INTRODUCTION

Micromachining technology combined with semiconductor processing provides an output that can be used to sense mechanical, optical, magnetic, chemical, biological, and other phenomena. These semiconductor sensors are, in most cases, a considerable improvement over the previous mechanical or other counterparts. However, as stand-alone components, they are far from the ideal characteristics that are desired for most measurements. In this chapter, the output from piezoresistive, capacitive, and piezoelectric sensors will be analyzed. Brief consideration will be given for other sensing techniques. The actual level of these signals and the various parameters used to specify the sensors' performance will be discussed.

3.2 SENSOR OUTPUT CHARACTERISTICS

The sensing technique used for a particular measurement can vary considerably depending on the range of the measurand, the accuracy required, the environmental considerations that impact packaging and reliability of the sensor, the dynamic nature of signal, and the effect of other inputs on the measurand. The environmental considerations include operating temperature, chemical exposure, and media compatibility. The measurement systems also apply constraints. Factors such as signal conditioning, signal transmission, data display, operating life, servicing/calibration, impedance of sensor, impedance of system, supply voltage, frequency response, and filtering may make one solution preferable over another. In some cases, the sensor's design can make more than one sensing technology acceptable. This depends on the design and manufacturing capability of the sensor supplier. For example, ceramic and semiconductor sensors using capacitive or piezoresistive technology frequently compete for a particular application. Also, optoelectronic, Hall effect, magnetoresistive, and even inductive sensors can measure displacement or velocity. However, once a technology has been accepted for a particular measurement, displacement by an alternate technology is quite difficult.

Determining the proper sensing technology for a particular application begins with understanding the fundamental design principle of the sensor and the specifications that the manufacturer guarantees. Semiconductor sensors have defined new terms and allow alternative design methodology for sensors based on the micrometer and nanometer scale in which they operate. Piezoresistive pressure sensor examples will be used to explain some key sensor parameters and a nontraditional design approach.

3.2.1 Wheatstone Bridge

The change in the resistance of a material when it is mechanically stressed is called "piezoresistivity." Strain-gauge pressure sensors convert the change in resistance in four (sometimes only one or two) arms of a Wheatstone bridge. The output voltage of a four-element Wheatstone bridge (Figure 3.1), is given by

$$E_\text{o} = \frac{E\Delta R}{R} \tag{3.1}$$

where E_o is the output voltage, E is the applied voltage, R is the resistance of all bridge arms, and ΔR is the change in resistance due to an applied pressure [1]. Additional variable resistive elements are typically added to adjust the zero offset, calibrate sensitivity, and provide temperature compensation. The Wheatstone bridge can be operated in a constant voltage or constant-current mode. The constant-voltage mode is more common since it is easier to generate a controlled voltage source. However, the constant-current mode is also used and provides temperature compensation that is independent of the supply voltage without reducing the voltage supplied to the bridge.

Different approaches to piezoresistive strain gauges range from traditional bonded and unbonded to the newest integrated silicon pressure sensors. Pressure applied to the diaphragm produces a change in the dimension of the diaphragm, an increase in the length of the gauge, and a change in its resistance ($R = \rho L/A$). The change in length per unit length is called "strain." The sensitivity of a strain gauge is indicated by gauge factor (GF), which is defined by

$$\frac{\Delta R}{R} = K \cdot \frac{\Delta l}{l} \tag{3.2}$$

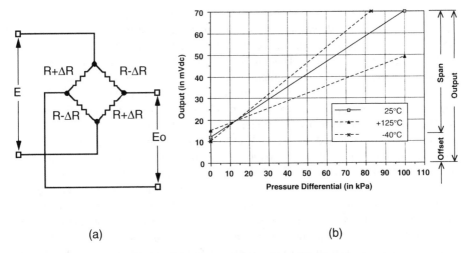

Figure 3.1 (a) Wheatstone bridge, and (b) output curve for piezoresistive pressure sensor.

where $K = \pi \cdot E$ = gauge factor, where π = the piezocoefficient and E = the modulus of elasticity for the material.

The K for metals is 2; for silicon it is approximately 100 to 200, depending on the doping level [2]. In contrast to metals, the change in resistance is a second-order effect for silicon. The primary effect is the change in conductivity of the semiconductor material due to its dependence on mechanical stress.

The GF for a strain gauge is improved considerably (to about 150) by using a silicon strain gauge. However, besides the conventional Wheatstone bridge, silicon processing techniques and the relative size of piezoresistive elements in silicon enable the design of a unique piezoresistive sensor. This alternate design demonstrates one of the different possibilities that semiconductor technology brings to sensing.

3.2.2 Piezoresistivity in Silicon

The analytic description of the piezoresistive effect in cubic silicon can be reduced to two equations that demonstrate the first-order effects [3].

$$\Delta\mathbf{E}_1 = \rho_0 \cdot I_1(\pi_{11}\mathbf{X}_1 + \pi_{12}\mathbf{X}_2) \tag{3.3}$$

$$\Delta\mathbf{E}_2 = \rho_0 \cdot I_2 \cdot \pi_{44}\mathbf{X}_6 \tag{3.4}$$

where $\Delta\mathbf{E}_1$ and $\Delta\mathbf{E}_2$ are electric field flux density, ρ_0 is the unstressed bulk resistivity of silicon, the I's are the excitation current density, the π's are piezoresistive coefficients, and the $\mathbf{X}$'s are stress tensors due to the applied force.

The effect described by (3.3) is utilized in a silicon pressure transducer of the Wheatstone bridge type. Regardless of whether the sensor designer chooses N-type or P-type layers for the diffused sensing element, the piezoresistive coefficients π_{11} and π_{12} of (3.3) will have opposite signs. This implies that through careful placement, orientation with respect to the proper crystallographic axis, and a sufficiently large aspect ratio for the resistors themselves, it is possible to fabricate resistors on the same diaphragm that both increase and decrease respectively from their nominal values with the application of stress.

The effect described by (3.4) is typically neglected as a parasitic in the design of a Wheatstone bridge device. A closer look at this equation reveals that the incremental electrical field flux density, ΔE_2, due to the applied stress, X_6, is monotonically increasing for increasing X_6. In fact, (3.4) predicts an extremely linear output since it depends on only one piezoresistive coefficient and one applied stress. Furthermore, the incremental electric field can be measured by a single stress-sensitive element. This is the theoretical basis for the design of the transverse voltage or shear stress piezoresistive strain gauge.

Figure 3.2 shows the construction of a device that optimizes the piezoresistive effect of (3.4). The diaphragm is anisotropically etched from a silicon substrate. The piezoresistive element is a single, four-terminal strain gauge that is located at the midpoint of the edge of the square diaphragm at an angle of 45 deg. The orientation of 45 deg and location at the center of the edge of the diaphragm maximizes the sensitivity to shear stress (X_6) and the shear stress being sensed by the transducer by maximizing the piezoresistive coefficient, π_{44}.

Basic uncompensated sensor element
— Top View

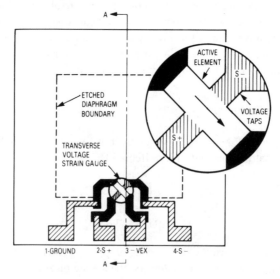

Figure 3.2   Shear stress strain gauge.

Excitation current is passed longitudinally through the resistor (pins 1 and 3) and the pressure that stresses the diaphragm is applied at a right angle to the current flow. The stress establishes a transverse electric field in the resistor that is sensed as an output voltage at pins 2 and 4, which are taps located at the midpoint of the resistor. The single-element shear stress strain gauge can be viewed as the mechanical analog of a Hall effect device.

Using a single element eliminates the need to closely match the four stress- and temperature-sensitive resistors on the Wheatstone bridge designs while greatly simplifying the additional circuitry necessary to accomplish calibration and temperature compensation. The offset does not depend on matched resistors, but on how well the transverse voltage taps are aligned. This alignment is accomplished in a single photolithographic step, making it easy to control, and is only positive, simplifying schemes to zero the offset. The temperature coefficient of the offset is small (nominally $\pm 15 \, \mu V/°C$) since multiple resistors and temperature coefficients do not have to be matched. By using proper doping levels, the temperature dependence of full- scale span (the difference between full-scale output and offset) can be carefully controlled and therefore compensated without the requirement of characterizing each device over the temperature range.

3.2.3 Semiconductor Sensor Definitions

The batch-processing techniques used by semiconductor manufacturers are ideally suited for making high-volume, low-cost sensors. One limitation of batch-processed parts, however, is that certain parameters are not precisely specified, but only listed as "typical" on the manufacturer's data sheet. In many cases, meeting fixed limits on all parameters can drive costs up and offset the benefits of high-volume batch processing [4].

Use of typical specifications is not necessarily a drawback, however. Uniformity within each wafer lot is a specific strength of semiconductor processing. Nevertheless, this factor must be considered to avoid problems in volume production.

The term "typical specification" generally has several implications in the semiconductor industry. Usually, it indicates a parameter that has been characterized during the design phase and represents the mean value for the manufacturing process. Certain manufacturers also add that a ± 3 sigma value is the total spread that users can expect on typical specified parameters. This spread cannot be assumed, however, because "typical" means that limits are not normally attached to the manufacturer's quality assurance program.

In many cases, "typical" is used to indicate that the measurement process contributes more to the inaccuracy of the reading than to the actual variation from unit to unit. This practice shows why users must evaluate the parameter specified as typical. Pressure hysteresis in silicon diaphragms, for example, is essentially nonexistent and should not cause great concern if specified as typical. On the other hand, the variation over temperature of span and offset compensation could be significant and requires further investigation if listed as a typical specification.

Many parameters on sensor data sheets are specified as $\pm \%$ of full scale (output or span). This designation is the key to understanding other specifications. Full-scale output is the output at rated pressure including the zero offset. Full-scale span is the difference between the output with no pressure applied (offset) and the output at rated pressure. For a given device (i.e., Motorola's MPX100), a 0.25% full-scale span rating on linearity can range from 0.113 mV for a low-output device (45 mV) to 0.225 mV for a device with the highest output (90 mV).

Sensitivity is the span divided by the operating pressure range. The sensitivities for devices made by various manufacturers are frequently compared. The pressure ranges and the supply voltage must be considered for an accurate comparison. This procedure often reveals interesting information that may affect a design decision, especially if the linearity at a given sensitivity is also taken into account.

The linearity or deviation from a straight-line relationship is expressed as a percent of full- scale output (or span) and is also subject to different test methods. The least-squares, best-fit technique is the commonly accepted choice for establishing linearity. However, it requires several measurement points to provide acceptable results.

A simpler measurement is the end-point method, which requires only three points: zero, mid scale, and full scale. This method lends itself to high-volume testing. Because it always yields a value that is about double the magnitude of a least-squares method, it is frequently more appropriate for error budget calculations. Any comparison of competitive data sheet values for linearity should take into account the method used to determine this parameter.

The effect of temperature on offset and span (and sensitivity) is one of the most critical, and frequently most troublesome, aspects of using semiconductor pressure sensors. Products that have no temperature compensation can be used with little effort for a narrow temperature range such as 25°C±15°C or less, and accuracy requirements of a few percent. Certain sensors, such as Motorola's MPX family, can be temperature compensated to 1% to 3% accuracy over a 0°C to 85°C temperature range without exercising the sensor over temperature. Ease of compensating for temperature is an important factor in evaluating a pressure sensor.

To accurately compare various products, similar units must be used. Full-scale value, temperature range, and actual limits (specified as ±% F.S., ±mV, or ±%F.S./°C) can vary from one manufacturer to another. The conversion of these units to mV/°C can provide a good comparison. For example, a calibrated pressure sensor, such as Motorola's MPX2000 family with 40 mV output and ±1%F.S. variation over 0 to 85°C, has a maximum window of 9.4 mV/°C. This type of specification can indicate improved performance over reduced temperature spans. However, a user must know actual curve shapes within the temperature compensation window to have confidence in this value.

Three other terms on sensor data sheets are often a source of confusion: static accuracy, ratiometricity, and overpressure. Frequently, manufacturers lump together linearity, pressure hysteresis, and repeatability as static accuracy. Sometimes this characteristic, simply stated as the accuracy, is indicated as a single parameter on the data sheet by some manufacturers.

Ratiometricity is one of the new terms that semiconductor manufacturers have brought to pressure sensing. Although output is given at a specific voltage rating, lower or higher voltage supplies can be used (within the maximum rating of the device) with constant-voltage source sensors. The output in this instance varies as the ratio of source voltage to the manufacturer's rated voltage. A constant voltage is required for proper operation, a fact that should not be overlooked when considering the total circuit requirements.

The term "overpressure" takes on a different meaning with semiconductor pressure sensors. Even though the silicon diaphragm can be approximately 0.001-in thick, a device rated at 30 psi can withstand 200 to 300 psi without damage. Most manufacturers specify overpressure conservatively at two or three times the rated pressure.

Obviously, no damage should occur when the sensor operates within the overpressure rating. However, readings made above the normal range are another matter. Normally, linearity starts to degrade and could fall outside the specified rating if the sensor is operated above its rated pressure. If linearity were only marginally acceptable to begin with, overscale readings could cause trouble. However, this is not the case for many applications.

The normal mind-set for making pressure readings with a pressure gauge is to avoid operating at full scale for best accuracy. Full-scale operation can also pin the gauge needle, resulting in miscalibration or damage. However, operating a semiconductor sensor at full scale usually provides better accuracy than a midscale reading. This is especially true for units measured for end-point linearity because output is accurately measured to specification at full scale. Slight excursions over the pressure rating do not significantly degrade linearity. However, if the output saturates above the rated pressure or the digital gauge's operating range is exceeded, a higher pressure input will not produce a higher output reading.

3.2.4 Static versus Dynamic Operation

Response time for semiconductor pressure sensors is reduced by the protective gels or coatings that protect the active surface from the pressure media. Also, isolating the semiconductor by stainless steel diaphragms and oil-filled chambers can reduce response time. However, response time within 1 ms is typically achieved when the sensor is exposed to a full-scale pressure excursion.

Measuring systems that operate at 6,000 rpm require sensors to operate at frequencies above 100 Hz. Frequently, a higher frequency range or higher frequency signal components are also of interest, as in acceleration and vibration sensors. However, lower frequency sensors tend to have lower noise floors. The lower noise floor increases the sensor's dynamic range and may be more important to the application than higher frequency capability [5].

3.3 OTHER SENSING TECHNOLOGIES

Piezoresistive sensing is the most common micromachined sensor since its output signal is predictable and easy to signal condition. Other techniques are gaining in popularity, especially as sensing techniques for smart sensors based on the capability of electronics circuitry to handle the signal conditioning. A brief review focusing on the output of current semiconductor sensing techniques will be given.

3.3.1 Capacitive Sensing

Capacitive sensors typically have one plate that is fixed and one that moves as a result of the applied measurand. The nominal capacitance is $C = A\varepsilon/d$, where A is the area of the plate, ε is the dielectric constant, and d is the distance between the plates. Two common capacitive pressure sensors used in automotive applications are based on silicon and ceramic capacitors. A differential capacitor structure can also be used for acceleration. Figure 3.3 shows the surface micromachined capacitor and the resulting output [6].

Silicon technology combined with surface micromachining has allowed the capacitance between interdigitated fingers to measure acceleration and other inputs. The value of the nominal capacitance is in the range of 100 fF to 1 pF and the variation in capacitance is in femtofarads. Circuitry, either integrated on the same silicon chip or in the same package (that takes into account the parasitic capacitance effect of packaging), can use this level of signal to provide a useful output for a control system [7].

3.3.2 Piezoelectric Sensing

A piezoelectric sensor produces a change in electrical charge when a force is applied across the face of a crystal, ceramic, or piezoelectric film. The inherent ability to sense vibration and necessity for high-impedance circuitry are taken into account in the design of modern piezocrystal sensors. Transducers are constructed with rigid multiple plates and a cultured-quartz sensing element, which contains an integral accelerometer to minimize vibration sensitivity and suppress resonances.

Lead zirconate titanate (PZT) ceramic is used to construct biomorph transducers that sense motion, vibration, or acceleration and can be activated by an applied voltage. The biomorph consists of two layers of different PZT formulations that are bonded together in rectangular strips or washer shapes. The rectangular shape can be mounted as a cantilever. Motion perpendicular to the surface generates a voltage [8].

More recently, piezofilm sensors, which produce an output voltage when they are deflected, provide a method for inexpensive measurements. Figure 3.4(a) shows the construction of a piezoelectric film sensing element [9]. The polymer film generates a charge when it is deformed and also exhibits mechanical motion when a charge is applied to it. For a vehicle detector, the sensor is mounted in an aluminum channel filled with

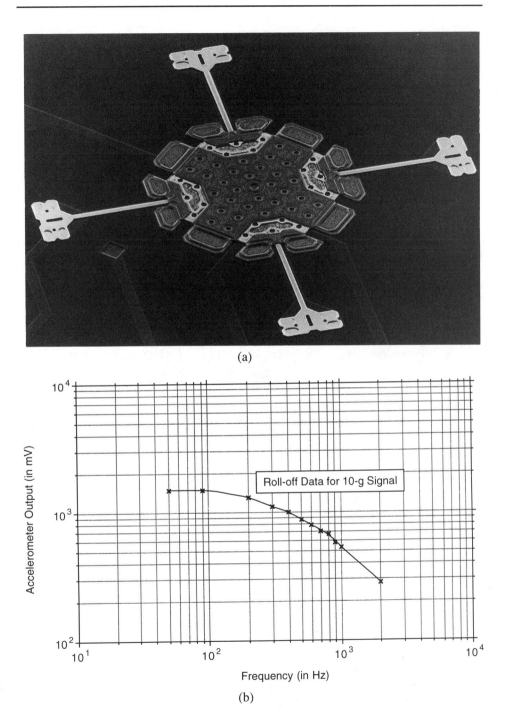

(a)

(b)

Figure 3.3 (a) Differential capacitor structure in accelerometer, and (b) output.

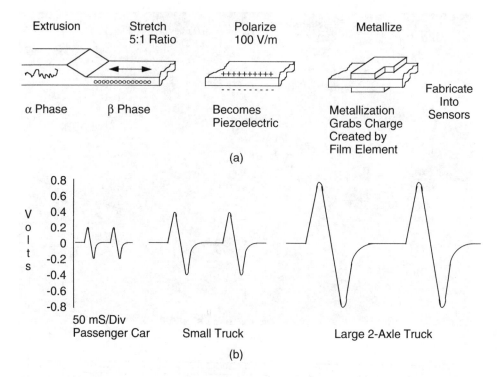

Figure 3.4 (a) A Kynar piezoelectric film is polarized in an intense electric field of approximately 100 V/μm and metalized to create a transducer; (b) the output of a vehicle sensor in a voltage mode is a voltage proportional to weight and speed of the vehicle. (*After:* [9].)

polyurethane and embedded in the pavement. The output of this sensor for three different types of vehicles is shown in Figure 3.4(b). In the voltage mode, the voltage is influenced by both the weight and the speed of the vehicle passing over it. A positive output occurs when the tire compresses the film. The negative output results from expansion after the tire has passed over the sensor.

Surface micromachining techniques have been combined with piezoelectric thin-film materials, such as zinc oxide, to produce a semiconductor piezoelectric pressure sensor. One design has a sensitivity of 0.36 mV/fbar at 1.4 kHz [10]. Low-level acoustic measurements are a potential application for this technology.

3.3.3 Hall Effect

A vertical Hall effect structure has been designed for the detection of magnetic fields oriented parallel to the plane of the chip. Bulk micromachining was used to achieve higher sensitivity than a device without micromachining. The micromachined unit had an output of 70 mV for an applied magnetic field of 400 mtesla, which was almost five times the sensitivity of the unetched unit [11].

3.3.4 Chemical Sensors

A metal oxide chemical sensor's resistivity changes depending on the reducing or oxidizing nature of the gaseous environment around the sensor. Figure 3.5 shows sensitivity data for H_2 and CO in air for one design [12]. Conductance of over $100(10^{-7})\ \Omega^{-1}$ for H_2 and almost $32(10^{-7})\ \Omega^{-1}$ for CO were measured with the sensor exposed to concentration of 2,000 ppm of each gas in separate measurements and operating at a minimum temperature of 250°C.

3.3.5 Improving Sensor Characteristics

The low-level output of the sensors described in this section and the design parameters discussed in Section 3.2.3 are among the characteristics that must be considered to use a semiconductor sensor in a particular application. Table 3.1 shows a summary of sensor

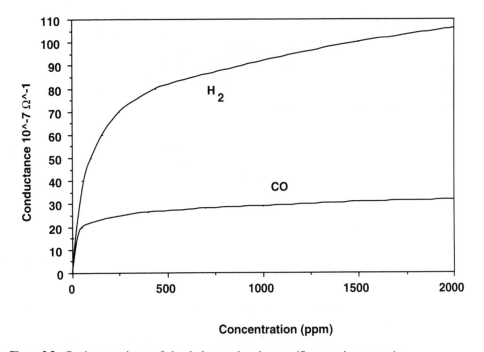

Figure 3.5 Conductance change of chemical sensor based on specific gas and concentration.

Table 3.1 Common Undesirable Characteristics of Semiconductor Sensors

Characteristic	Sensor Design	Sensor Interface	MCU/DSP
Nonlinearity	Consistent		Reduce
Drift	Minimize		Compensate
Offset		Calibrate	Calibrate/reduce
Time dependence of offset	Minimize		Autozero
Time dependence of sensitivity			Autorange
Nonrepeatability	Reduce		
Cross-sensitivity to temperature and strain		Calibrate	Store value & correct
Hysteresis	Predictable		
Low resolution	Increase	Amplify	
Low sensitivity	Increase	Amplify	
Unsuitable output impedance		Buffer	
Self-heating	Increase Z		PWM technique
Unsuitable frequency response	Modify	Filter	
Temperature dependence of offset			Store value & correct
Temperature dependence of sensitivity			Store value & correct

characteristics and the design area that is commonly used to improve these characteristics [13]. Improved sensor performance can result when the design of the sensor and capability of subsequent components that achieve a smart sensor are taken into account in the overall design of the smart sensor. Digital logic provided by either an MCU or DSP plays an vital role in smart sensing and in improving the sensor's performance.

3.4 DIGITAL OUTPUT SENSORS

A sensor that directly interfaces to an MCU without requiring analog-to-digital conversion simplifies system design and reduces the cost of the MCU. Today's sensors accomplish this with additional circuitry. However, the industry quest is a sensor or sensor family with an inherent digital output. "Inherent" may be achieved by on-chip or external signal conditioning, but they must be considerably lower cost and more accurate than units using current techniques.

3.4.1 Incremental Optical Encoders

Structures that contain a light source and photodetector provide a digital means of measuring displacement or velocity when an alternating opaque and translucent grid is passed between them. As shown in Figure 3.6, a macro-sized example of uniformly spaced apertures on a wheel allows logic circuitry to count the number of pulses in a given

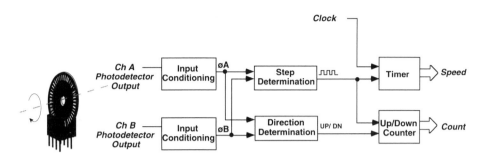

Figure 3.6 Incremental optical encoder.

timeframe to determine shaft velocity or angular displacement [14]. Linear measurements of displacement and velocity are also possible.

Basic presence sensing can be accomplished through the use of a single optical channel (or emitter-detector pair). Speed and incremental position sensing must use two channels. The most commonly used approach is quadrature sensing, where the relative positions of the output signals from two optical channels are compared. This comparison provides the direction information, and either of the individual channels gives the transition signal to derive either count or speed information. A typical application may use the direction output as the up/down input and either channel A or B as the count input for a common up/down counter. The counter in an incremental system will increment or decrement as required to maintain a relative position or count output.

Quadrature direction sensing requires that the two optical channels create electrical transition signals that are out of phase with each other by 90° nominally. When the code wheel is spinning, the two electrical output signals, A and B, will be 90° out of phase, because the wheel windows are 90° out of phase with the two sensor channels. The interpretation of whether signal A leading B is clockwise or counterclockwise is a matter of choice. Figure 3.7 shows the ideal quadrature outputs. The two waveforms are in four equal quadrants, exactly 90° out of phase with each other. The detector is ''ON'' when light is present and turns ''OFF'' when the web blocks the beam.

3.4.2 Digital Techniques

A study of available digital techniques has been reported [13]. Sensing techniques based on a resonant structure or a periodical geometric structure are potentially the most direct approaches to digital sensing. However, other techniques can be used as well.

Micromachined silicon can act as a resonant structure if it is designed with a membrane or tuning fork. Electric activation is achieved by using a piezoelectric film such as ZnO on the micromachined structure. Most resonant structures are implemented in pairs. One element is not interfaced to the input and acts as a reference. Comparison of the output

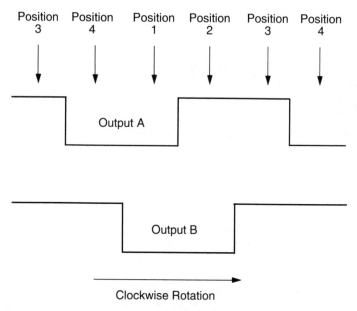

Figure 3.7 Quadrature detection.

frequency of the sensing element to the reference element reduces the influence of unwanted parameters. This technique is commonly used in chemical sensors using surface acoustical wave (SAW) delay-line oscillators.

Other approaches for digital output sensors include electrical oscillator-based (EOB) sensors and stochastic analog-to-digital (SAD) converters [13]. An EOB sensor is designed to generate a periodic current or voltage signal when it is subjected to a measurand. Current-to-frequency or voltage-to-frequency conversion provides a digital signal. The current-to-frequency technique has been demonstrated on a piezoresistive silicon pressure sensor. The frequency changed from 200 to 230 kHz with a pressure change from 0 to 750 millimeters mercury (mmHg).

A different approach for an EOB sensor was used with a capacitive sensor. Two equal but opposite current sources applied a square wave to a capacitive silicon pressure sensor. The zero-pressure frequency was 155 kHz and the sensitivity was 25 kHz/450 mmHg for this design.

A ring oscillator has also been used for EOB sensors. An odd number of I^2L (integrated- injected logic) gates connected in a ring on a piezoresistive pressure sensor provided the ring oscillator. Sensitivities of up to 10.6 kHz/bar were obtained with a ring oscillator frequency of 667 kHz.

SAD converters employ the noise in flip-flop circuits to generate a random signal and use the flip-flop as a comparator. A piezoresistive sensor element in the flip-flop circuit is one way to implement this approach for a sensor. The number of ones and zeros is a direct measure of the strain in a pressure sensor and a simple counting procedure for an MCU. A flip-flop sensor that can measure stresses as small as 8 kilo Pascals (kPa) in a silicon cantilever beam has been reported.

A final digital example is provided by arrays of sensing elements [15]. A pressure switch with four switch points was fabricated using silicon fusion bonding. Pressure-switch points of 1/4, 1/2, 3/4, and 1 atmosphere were designed using controlled thinning of the diaphragm. The switches close electrical contacts when their desired pressure threshold is exceeded. Table 3.2 shows the truth table for the design. The outputs can be directly applied to a logic control.

3.5 NOISE/INTERFERENCE ASPECTS

The general model for a transducer shown previously in Figure 1.2 in Chapter 1 indicated the interference sources that affect the low-level output of the sensor. An actual sensor signal combined with noise and interference from other sources, such as temperature, humidity, and/or vibration, can be a major problem when dynamic signals are measured. Compared to static signals where filtering can be used to minimize noise, dynamic measurements can pose a challenge [16].

3.6 LOW-POWER, LOW-VOLTAGE SENSORS

Portable sensing applications, including portable data loggers and data acquisition systems, require sensors to operate at low power levels and low battery supply voltages. For low-power pressure measurements, high-impedance pressure sensors and pulse width modulation (PWM) techniques have been developed to reduce power drain. Depending on the type of measurement (static or dynamic) and how frequently it must be measured, the sensing system can be in a sleep mode and wake up to make periodic readings that can be transmitted to distant recording instruments. This technique is useful in process controls, hazardous material monitoring systems, and a variety of data acquisition applications that would previously have been more time consuming, dangerous, prohibitively expensive, or too heavy with previous technology. Monitoring pressure is one of the more frequent

Table 3.2 Truth Table for Array of Four Pressure Switches

Pressure (P)	S1	S2	S3	S4	Out1	Out2	Out3	Out4
P < P1	Open	Open	Open	Open	0	0	0	0
P1 ≤ P < P2	Closed	Open	Open	Open	1	0	0	0
P2 ≤ P < P3	Closed	Closed	Open	Open	1	1	0	0
P3 ≤ P < P4	Closed	Closed	Closed	Open	1	1	1	0
P4 ≤ P	Closed	Closed	Closed	Closed	1	1	1	1

measurements that must be made in systems ranging from tire pressure on vehicles to leakage monitoring in tanks, blood pressure in portable healthcare monitors, and barometric pressure in weather measurements. The combination of higher impedance sensors with new design techniques, such as PWM input and other power management techniques, is allowing this area of portable equipment to increase.

3.6.1 Impedance

Many mechanical transducers have a low impedance. Strain gauges, for example, are typically 350Ω. This impedance has been achieved in semiconductor devices, especially where direct replacement of a mechanical unit is required. In semiconductor circuits, low impedance is desired for noise purposes. However, high impedance is required to minimize the current draw for portable applications, to use existing interface circuits, and to prevent loading on amplification stages [17]. For high-impedance pressure sensors, an input impedance in the range of approximately 5 kΩ is common. Output impedance for these devices is also approximately 5 kΩ. The sensor must be designed to achieve the results that the particular system requires.

3.7 AN ANALYSIS OF SENSITIVITY IMPROVEMENT

Low-pressure measurements (< 0.25 kPa or 0.036 psi) provide a good example of the problems that can occur and the solutions that have been developed to cope with the low-level signal. Low-pressure measurements are limited by existing silicon sensor designs. The sensitivity of a silicon diaphragm is directly related to the area and inversely related to the square of the thickness of the diaphragm.

3.7.1 Thin Diaphragm

One approach for increasing the sensitivity for low-pressure measurements is thinning the diaphragm. Typical thicknesses for silicon diaphragms range from 2 to 12 μm. However, a thin diaphragm can have unacceptable linearity. A number of researchers have investigated a variety of stress concentrators or bosses designed into the diaphragm structure to minimize the nonlinearity. The bosses provide a locally stiffer structure and limit the overall deflection of the diaphragm. A thinner diaphragm (4 fm), shallow resistor geometries in the submicron area, and advanced silicon micromachining have been used in a piezoresistive sensor to achieve a sensitivity of 50 mV/V/psi (7.3 mV/V/kPa) [18].

3.7.2 Increase Diaphragm Area

A second approach to low sensitivity involves increasing the sensitivity by increasing the size of the diaphragm. An area increase of over 2.5 times results in a 0.5-psi sensor with

7 mV/V/psi [19]. This larger area does not compromise ruggedness, but it does increase the cost of the sensor.

3.7.3 Combined Solution: Micromachining and Microelectronics

A more recent solution to increasing the sensitivity uses the combined capability of a low-pressure sensor and a microcontroller. Figure 3.8 shows the block diagram [20]. The microcontroller provides a signal to pulse the sensor with a higher supply voltage. The microcontroller also provides signal averaging for noise reduction and samples the supply voltage to reject sensor output variations due to power supply variations. An important factor for sensors making very low pressure measurements is stress isolation for the package. A piston-fit package for the unit used in this circuit allows 40 μV/Pa with a 15V source.

3.8 SUMMARY

This chapter has shown the distinct advantages that semiconductor sensors have over mechanical predecessors. However, these sensors require additional effort to obtain their full benefit in systems. Semiconductor sensors draw from previous mechanical sensors, but the physics of the micro scale can provide alternative solutions that are not possible at the macro level.

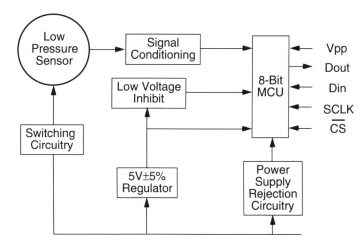

Figure 3.8 Low-pressure sensor combined with MCU to achieve higher pressure sensitivity.

REFERENCES

[1] Carr, J. J., *Sensors and Circuits*, Englewood Cliffs, New Jersey: Prentice-Hall, Inc., 1993.

[2] Shankland, E. P., "Piezoresistive Silicon Pressure Sensors," *Sensors*, Aug. 1991, pp. 22–26.

[3] Frank, R. K., and W. E. McCulley, "An Update on the Integration of Silicon Pressure Sensors," *Wescon/85 Integrated Sensor Technology Session Record 27*, San Francisco, CA, pp. 27/4-1–27/4-5.

[4] Frank, R., "Designing with Semiconductor Pressure Sensors," *Machine Design*, Nov. 21, 1985, pp. 103–107.

[5] Barrett, R., And F. Wicoxon, "Monitoring Vibration with Piezoelectric Sensors," *Sensors*, Aug. 1993, pp. 16–24.

[6] Ristic, L. et al., "A Capacitive Type Accelerometer with Self-Test Feature Based on a Double-Pinned Polysilicon Structure," *Digest of Technical Papers Transducers '93*, Yokohama, Japan, June 7–10, 1993, pp. 810–813.

[7] Payne, R. S., and K. A. Dinsmore, "Surface Micromachined Accelerometer: A Technology Update," SAE P-242 *Sensors and Actuators*, 1991, pp. 127–135.

[8] Matroc, M., "An Update on Piezoelectric Ceramic Transducers," *Electronic Design*, Oct. 26, 1989, p. 119.

[9] Halvorsen, D. L., "Piezoelectric Polymer Traffic Sensors and the Smart Highway," *Sensors*, Aug. 1992, pp. 10–17.

[10] Schiller, P., D. L. Polla, and M. Ghezzo, "Surface Micromachined Piezoelectric Pressure Sensors," *Technical Digest of IEEE Solid-State Sensor and Actuator Workshop*, Hilton Head, SC, 1990, pp. 188–190.

[11] Paranjape, M., Lj. Ristic, and W. Allegretto, "Simulation, Design and Fabrication of a Vertical Hall Device for Two-Dimensional Magnetic Field Sensing," *Sensors and Materials*, Vol. 5, No. 2, 1993, pp. 91–101.

[12] Chemical Sensor Brochure, Microsens, Neuchatel, Switzerland.

[13] Middelhoek, S., P. J. French, H. J. Huijsing, and W. J. Lian, "Sensors with Digital or Frequency Output," *Sensors and Actuators* 15, Netherlands: Elsevier Sequoia, 1988, pp. 119–133.

[14] Cumberledge, W., R. Frank, and L. Hayes, "High Resolution Position Sensor for Motion Control System," *Proc. of PCIM '91*, Universal City, CA, Sept. 22–27, 1991, pp. 149–157.

[15] Ismail, M. S., and R. W. Bower, "Digital Pressure-Switch Array with Aligned Silicon Fusion Bonding," *Journal of Micromechanics and Microengineering*, Vol. 1, No. 4, Dec. 1991, pp. 231–236.

[16] Cigoy, D. et al., "Low Noise Cable Testing and Qualification for Sensor Applications," *Proc. of Sensors Expo*, Cleveland, OH, Sept. 20–22, 1994, pp. 575–596.

[17] Motorola Pressure Sensor Device Data Book, DL200/D Rev. 1, 1994.

[18] Bryzek, B., J. R. Mallon, and K. Petersen, "Silicon Low Pressure Sensors Address HVAC Applications," *Sensors*, March 1990, pp. 30–34.

[19] Hughes, B., "Sensing Pressure Below 0.5 psi," *Sensors Expo Proc.*, 1990, pp. 104A1–104A5.

[20] Baum, J., "Low-Pressure Smart Sensing Solution with Serial Communications Interface," *Proc. of Sensor Expo*, Boston, MA, May 16–18, 1995, pp. 251–261.

Chapter 4

Getting Sensor Information into the MCU

"Someday we'll find it—the rainbow connection."
—Kermit the Frog from Jim Henson's Muppets

4.1 INTRODUCTION

The sensor signal is only the first step towards a sensor that will ultimately provide an input to a control system. Signal conditioning the output from the sensor is as important as choosing the proper sensing technology. Most transducer elements require amplification as well as offset and full-scale output calibration and compensation for secondary parameters such as temperature. Whether this signal conditioning is performed as an integral part of the purchased sensor or performed by the user, the accuracy of the measurement will ultimately be determined by the combination of the sensor's characteristics and the additional circuitry. The signal-conditioning circuitry that is required is a function of the sensor itself and will vary considerably depending upon whether the sensor is capacitive, piezoresistive, piezoelectric, and so on.

Once a high-level analog signal is available, it must be converted to a digital format for use in a digital control system. This analog to digital converter (A/D converter or frequently ADC) can be an integral part of the digital controller or it may be an intermediate standalone unit. Different approaches are used for A/D converters and the theoretical accuracy (resolution), indicated by the number of bits, is not always achieved in the application.

This chapter will address signal conditioning and A/D conversion for sensors. Examples will be shown of available integrated circuits (ICs) specifically designed to simplify the task of signal conditioning a broad range of sensors and applications circuits developed for piezoresistive pressure sensors with specific performance characteristics.

4.2 AMPLIFICATION AND SIGNAL CONDITIONING

Micromachining is used to manufacture a diaphragm or beam thickness to nominal targets. Microelectronics is used to provide the precision for semiconductor sensors. Accuracy combined with ease of interface, cost, power consumption, printed circuit board space, and power supply voltage can be among the considerations when selecting a signal-conditioning IC. The semiconductor technologies used for the modular amplifiers and ICs in this section have a major impact on these criteria.

Sensor signal-conditioning circuits are based on two fundamental technologies: bipolar and complementary metal oxide semiconductor (CMOS). However, these two technologies have a vast number of derivatives. The requirements of the application should be used to determine which technology or derivative is appropriate. A performance comparison between bipolar and CMOS for signal-conditioning circuits is shown in Table 4.1 [1]. These two technologies can be combined to obtain BiMOS and the best features of bipolar and CMOS at the expense of a more complex process—going from a 10-mask bipolar process to one requiring 14 or 15 mask steps. The BiMOS process has lower yields because of the added processing complexity, so applications must utilize the performance improvements that it provides.

A 5V supply is commonly used for microcontrol units (MCUs) and digital signal processing (DSP). Consequently, common signal-conditioned outputs for sensors are 0.5 to 4.5V or 0.25 to 4.75V. Other standard industrial outputs are 1 to 6V, 1 to 5V, and 0 to 6V. The supply voltage for these units can range from 7 to 30V.

Signal conditioners for sensor outputs range from basic low-gain dc amplifiers to specialized amplifiers such as phase-sensitive demodulators [2]. Linear, logarithmic, high-gain, or dc-bridge amplifiers all fall within the following categories: differential amplifier, ac-coupled amplifier, chopper-stabilized amplifier, carrier amplifier, dc-bridge amplifier, and ac-level amplifier. A number of specialized amplifiers are also used for sensors, including log-linear amplifiers, frequency to voltage converters, integrator amplifiers, and differentiator amplifiers. These packaged solutions take the low-level output from sensors and add computational processes to free the digital controller from performing time-

Table 4.1 CMOS to Bipolar Comparison

Circuit Element	Bipolar	CMOS
Op amp	Very good	Fair
Analog switch	Poor	Very good
Comparator	Very good	Fair
A/D converter	Good	Good
Reference	Very good	Poor
Mirror	Good	Very good
Regulator	Very good	Poor
Active filter	Good	Very good
Logic	Poor	Very good
Power amplifier	Good	Good
< 3V operation	Very good	Good

consuming functions. Some amplifiers work with either ac or dc excitation and others are designed specifically for ac or dc operation.

In any sensor amplifier circuit, circulating currents in the ground path between a sensor and the measurement point can generate a common-mode voltage. The voltage appears simultaneously and in-phase at both amplifier input terminals. This voltage can either produce an error in the measurement or a catastrophic failure. The common-mode rejection ratio (CMRR), expressed in decibels, is a measure of an amplifier's ability to avoid this problem. A differential amplifier rejects common-mode voltage and amplifies only the difference across its input terminals.

4.2.1 Instrumentation Amplifiers

Several ICs have been designed to provide amplification for applications including sensors. Instrumentation amplifiers are used to provide a differential-to-single ended conversion and amplify the low-level signal from a sensor. Ratiometricity, or an output that varies linearly with the supply voltage, is an important consideration for sensors and signal-conditioning circuitry used to amplify their output. The impedance of the sensor, temperature range, and desired measurement accuracy are important factors in deciding the type of signal-conditioning circuitry that is required.

The instrumentation amplifier's dedicated differential-input gain block differentiates it from an ordinary operational amplifier (op amp). Although the signal level for the sensor may be only a few millivolts, it may be superimposed on a common-mode signal of several volts. A high common-mode rejection ratio keeps the common-mode voltage fluctuations from causing errors in the output [3]. A simple circuit using a standard instrumentation amplifier with a gain of 10,100, CMRR of 100 dB and a bandwidth of 33 kHz is shown in Figure 4.1 [4]. The pressure sensor has passive calibration for zero offset, full-scale span, and temperature compensation, but no additional amplification. The bridge current in the circuit is less than 3.6 mA with a 9V supply, and the output can be directly interfaced to an ADC of an MCU or DSP.

Another circuit that demonstrates the amplification and calibration for a pressure sensor is shown in Figure 4.2 [5]. This circuit has a controlled voltage for the bridge and amplifier supplied from voltage regulator U1. The supply voltage can range from 6.8V to 30V [6]. The low-cost interface amplifier provides a 0.5 to 4.5V output and can achieve an accuracy of ±5% by using 1% resistors. The MCU can increase this accuracy through software and avoid an analog trim in the amplifier.

4.2.2 Sleep-Mode™ Operational Amplifier

Integrated circuits designed to optimize performance may provide new solutions for smart sensors. For example, in battery-powered applications the current drain and power consumption must be low to obtain optimum battery life. A bipolar operational amplifier has

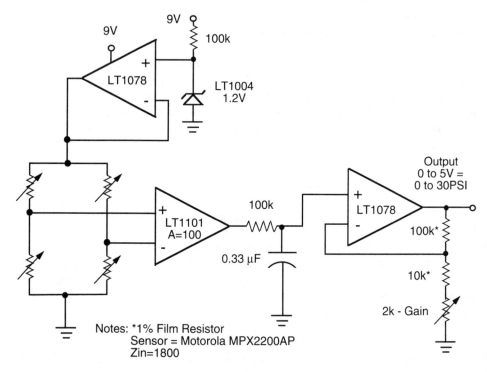

Figure 4.1 Pressure sensor circuit using an instrumentation amplifier.

been designed that conserves power in portable and other low-power applications but can handle higher current drain for improved performance [7]. The Sleep-Mode operational amplifier consumes only 70 μA maximum (40 μA typical) with 15V when it is asleep. In the awake mode, it has a typical gain-bandwidth product of 4.6 MHz, CMRR of 90 dB, and can drive 600 μA. The key elements of this IC are shown in Figure 4.3. In the sleep mode, the amplifier is active and waiting for an input signal from the sensor. When a signal is applied that sinks or sources 160 μA, the amplifier automatically switches to the awake mode for higher slew rate, increased gain bandwidth, and improved drive capability.

4.2.3 Rail-to-Rail™ Operational Amplifiers

Battery-powered circuits require system operation at degraded battery conditions that could be as low as 2.1V for two alkaline batteries. A bipolar IC has been designed that avoids the high threshold voltage of CMOS amplifiers and operates at voltages as low as ±0.9V [8]. The rail-to-rail design has an output stage that is current-boosted to deliver at least 50 mA to the load within 50 mV of the supply rails. The switchable NPN/PNP input stage

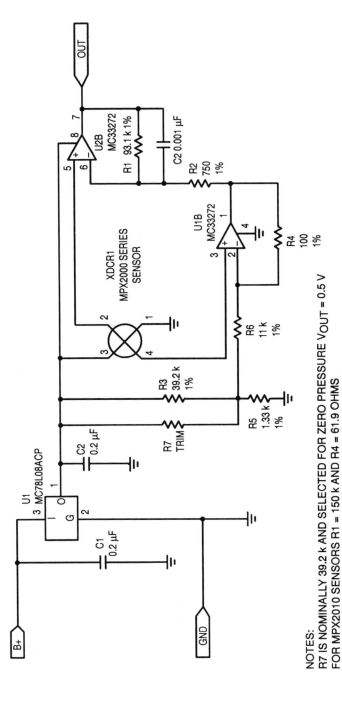

NOTES:
R7 IS NOMINALLY 39.2 k AND SELECTED FOR ZERO PRESSURE V_{OUT} = 0.5 V
FOR MPX2010 SENSORS R1 = 150 k AND R4 = 61.9 OHMS

Figure 4.2   Sensor-specific signal-conditioning circuit.

Simplified Block Diagram

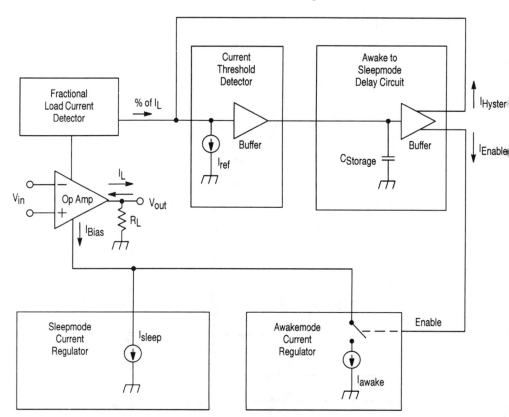

Figure 4.3 Block diagram of Sleep-Mode™ operational amplifier.

can be used as a voltage follower. Specialized circuits of this type can be useful in providing performance for sensor signal conditioning in specific applications.

The effect of rail-to-rail operation is shown in Figure 4.4 [9]. A CMOS process has difficulty achieving low-drift and low-noise requirements of high performance analog functions. For the amplifier to contribute no additional error to the ADC, the amplifier's signal-to-noise (S/N) ratio should be below the theoretical best-case dynamic range of the ADC. Depending on the amount of noise that is present, the combination of the drive amplifier and ADC will provide an effective number of bits of resolution. The A/D dynamic range is the full-scale value (high reference)-A/D low reference. The normal output from the sensor must be inside the A/D dynamic range to ensure proper operating headroom.

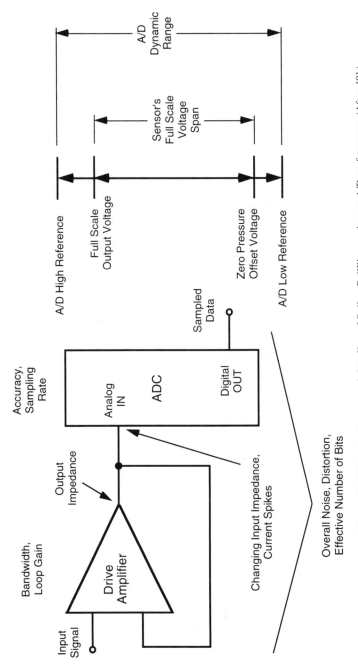

Figure 4.4 (a) Drive amplifier and ADC block diagram, and (b) effect of Rail-to-Rail™ operation on A/D performance. (*After:* [9].)

4.2.4 Switched-Capacitor Amplifier

CMOS is preferred for signal-conditioning sensors with capacitive output because of the high-input impedance of CMOS. Other advantages of CMOS are low power-supply current requirements and the ability to provide analog or digital outputs. Switched-capacitor techniques using CMOS circuitry have been developed to detect changes as small as 0.5 pF. A switched-capacitor filter is an analog-sampled data circuit. The components of a basic circuit are an operational amplifier, a capacitor, and an analog switch, all driven by a clock [10]. The filter's performance is determined by the ratio of the capacitor values and not the absolute values. Switching the capacitor makes it behave as a resistor with an effective resistance that is proportional to the switching rate. Figure 4.5 shows a switched-capacitor filter that has been developed for a surface-micromachined accelerometer with differential capacitive sensing [11]. This type of circuit minimizes effects of parasitic capacitance and has a very high input impedance for minimal signal loading.

4.2.5 Barometer Application Circuit

Sensor applications that use only a limited portion of the available output from a sensor can provide interesting challenges to sensor manufacturers and circuit designers. Barometric measurements, for example, use only a pressure variation of a few inches of mercury. In addition, this variation is superimposed on the altitude pressure reading, which can vary from 29.92 inches of mercury at sea level to 16.86 inches of mercury at 15,000 feet. A circuit that interfaces a temperature-compensated and calibrated 100-kPa (29.92 inches of mercury) absolute pressure sensor to a microcontroller has been designed based on a variation

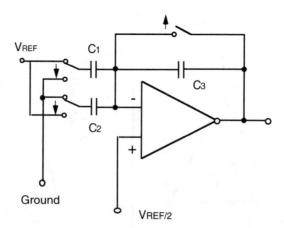

Figure 4.5 Switched-capacitor filter circuit.

of Figure 4.2 [12]. The circuit has a gain of 187 and obtains 0.1 inch Hg resolution when it is interfaced to an 8-bit MCU with an integral ADC.

4.2.6 A 4- to 20-mA Signal Transmitter

The use of monolithic two-wire transmitters can provide a low-cost easy to interface signal-conditioning solution for 4 to 20 mA current loops to transmit sensor readings in many industrial applications. The current loop avoids the problems of voltage drop when transmitting an analog signal over a long distance and the supply voltage can range from 9 to 40V. The circuitry in these devices can also compensate for nonlinearity in a resistance temperature detector (RTD) or resistance bridge sensor and therefore improve the overall performance of the sensor/signal-conditioning combination. Furthermore, the availability of additional features, such as a precision current source or 5V shunt regulator, can simplify the task of the circuit designer and allow use of these ICs for several types of sensors. Figure 4.6 shows a high-impedance pressure sensor interfaced with a transmitter IC [13]. With zero pressure applied, the output is 4 mA and with full pressure it is 20 mA. A 240-ohm resistor referenced to ground at the receiving end provides a 0.96 to 4.8V signal that can be interfaced to an ADC.

4.2.7 Schmitt Trigger

One final circuit to be considered is the Schmitt trigger. A Schmitt trigger turns the pulsed output from a sensor such as an opto detector or phototransistor into a pure digital signal. Figure 4.7 shows the opto input and output of the Schmitt trigger [14]. The lower and upper thresholds in the trigger remove the linear transition region between the on and off states. This hysteresis filters electrical noise that can cause the output to change state when it is close to the threshold of a digital IC input. The output of the Schmitt trigger can be used directly by digital logic circuits.

4.3 SEPARATE VERSUS INTEGRATED SIGNAL CONDITIONING

There is a worldwide effort [15, 16] in both academia and industry to integrate various sensors with electronics because of the potential advantages. These advantages include improved sensitivity, differential amplification for canceling parasitic effects such as temperature and pressure, temperature compensation circuits, multiplexing circuits, and A/D converting circuits [17]. The combination does not inherently offer lower cost or improved performance. However, selection of the proper sensing technique, application (especially high volume), and design criteria can provide an integrated sensor that is lower cost and more reliable than a thick-film multichip version.

Semiconductor sensors are subject, just as are other semiconductor devices, to the

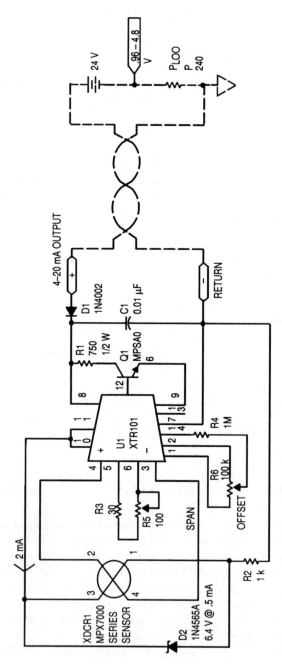

Figure 4.6 Pressure sensor with 4- to 20-mA transmitter.

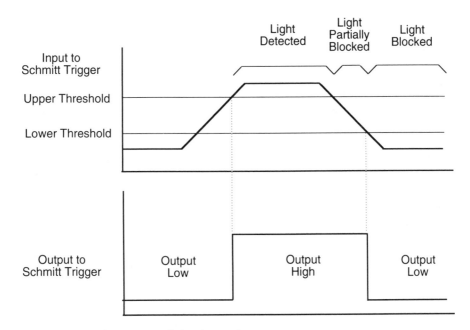

Figure 4.7   Schmitt trigger output and photodetector input.

paradox that (1) increased integration is inevitable with semiconductor technology for performance and cost reduction and (2) all devices cannot be integrated and obtain improved performance or cost reduction. Proper system partitioning is required to avoid components that are difficult to integrate, such as high-value capacitors, and components that can be obtained more cost-effectively by other approaches. Combining the sensor and additional circuitry can be beneficial in some applications and undesirable in others.

Precision outputs for integrated sensors can be obtained by laser trimming of zero TCR (temperature coefficient of resistance) thin-film metals at the wafer, die, or package level. Silicon resistors created by alloying silicon and aluminum can also be trimmed for improved accuracy. The ZIP-R™ trim process is a patented technique that can achieve essentially continuous adjustment by computer-controlled current pulses applied to these resistors either at the wafer level or after packaging [18].

Integrated electronic techniques are especially useful when only a small adjustment must be made. Active elements, such as op amps and transistors, are used to produce an amplified signal that can be easily interfaced to the rest of a sensing system. Improved precision can be obtained by characterizing devices over temperature and pressure (or acceleration or force) and storing correction algorithms in memory. Semiconductor sensing technology allows the integration of both the sensing element(s) and signal-conditioning circuitry into the same monolithic (silicon) structure. A number of factors determine which technology a manufacturer would choose when approaching a new design. The integration of passive elements and active elements with a piezoresistive pressure sensor will be used to demonstrate common approaches.

4.3.1 Integrated Passive Elements

A sensor with precisely trimmed offset, full-scale span, and temperature compensation can simplify the calibration procedure of the sensor and signal-conditioning circuitry. Figure 4.8 shows a pressure sensor with integrated resistors that are trimmed to meet specifications that are not possible in the semiconductor process flow. The low temperature coefficient of resistance (TCR) resistors are laser-trimmed with a 5-m diameter-pulsed laser beam [19]. Factors that affect the trim accuracy include the variability of the resistance of the deposited thin-film metal resistors, the annealing cycle for stress relief, and the sheet resistance. Process control allows the trimmed product to be extremely repeatable and provides high volume and high yield. This small amount of integration simplifies subsequent amplification circuitry as the circuits in Figure 4.1 and Figure 4.2 demonstrated. It also allows the sensor to be applied to a number of applications and with varied supply voltages.

4.3.2 Integrated Active Elements

Integrating the amplification and signal conditioning directly on the sensor can be accomplished using combined micromachining and microelectronics. The circuit of a fully signal-conditioned pressure sensor is shown in Figure 4.9(a). The additional circuitry is integrated on the sensor using available silicon area that is required to provide the mechanical support for the diaphragm [20]. The die size for the fully signal-conditioned unit is 145 mils by 130 mils. The diaphragm and piezoresistive element account for 20% of the die and the signal-conditioning circuitry is 80%, including the area required for wire bond pads. A quad operational amplifier that is used frequently for sensor output amplification is the LM324. The area of silicon in the LM324 is about 50 mils by 50 mils. The sensor without signal conditioning is a 120 mils by 120 mils chip. In this case, the electronics have been combined with the sensor with a very small increase in the total chip area because the mechanical support for the diaphragm has been utilized. Common pressure ranges and a standard interface to a 5V dc ADC allow the integrated approach to satisfy a number of application requirements and justify the cost of the integrated design. The integrated sensor easily interfaces to an MCU with an onboard ADC, as shown in Figure 4.9(b). The 50-pF capacitor and 51-kΩ resistor are recommended as a decoupling filter.

4.4 DIGITAL CONVERSION

Various A/D architectures are available for integration and interfacing with sensors. Conversion resolution, conversion accuracy, conversion speed or bandwidth, inherent

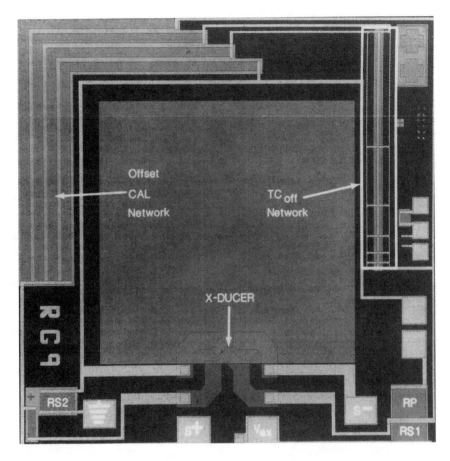

Figure 4.8 Photomicrograph of micromachined piezoresistive pressure sensor with integrated resistor network for zero offset, span, and temperature compensation.

system noise levels, and power consumption are all A/D converter tradeoffs. In assessing a converter architecture, it is important to consider all these aspects. For example, errors due to temperature, supply voltage, linearity, quantizing, and so forth may reduce the accuracy of an ADC by several bits when all error sources are considered. Also, bit accuracy alone may not be sufficient, especially if the sampling or conversion rate is incorrect for the sensor response under consideration [21]. Table 4.2 shows the quantizing errors and other parameters for 4 to 16-bit A/D conversion [22]. The quantization error (as a percentage of full-scale range) is $\pm 1/2 \cdot 1/(2^n - 1) \cdot 100$ which is also $\pm 1/2$ LSB. The resolution is the LSB (least significant bit) % full-scale/100. The theoretical rms S/N ratio for an N-bit ADC is calculated by the following equation:

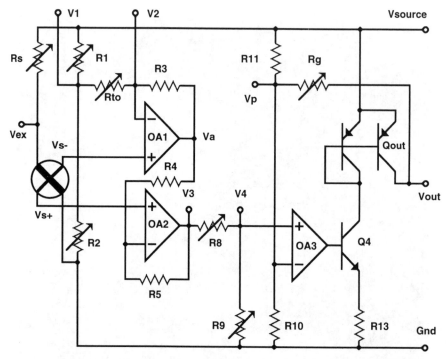

Figure 4.9 (a) Integrated piezoresistive pressure sensor circuit and interface to MCU, and (b) sensor interfaced to MCU.

$$S/N = 6.02 \cdot N + 1.76 \text{ dB} \qquad (4.1)$$

where N = number of bits.

4.4.1 A/D Converters

Common A/D conversion techniques include: single-slope (ramp-integrating), dual-slope integrating, tracking, successive approximation, folding (flash), and sigma-delta oversampled A/D converters. A comparison of the six types showing the relative silicon area required for each is shown in Table 4.3 [21]. The impact of hardware versus software driven successive approximation should be noted.

Most A/D converters can be classified into two groups based on the sampling rate, namely, Nyquist rate and oversampling converters. The Nyquist rate requires sampling the analog signals that have maximum frequencies slightly less than the Nyquist frequency, $f_N = f_s/2$, where f_s is the sampling frequency. However, input signals above the Nyquist frequency cannot be properly converted and create signal distortion or aliasing. A low-pass

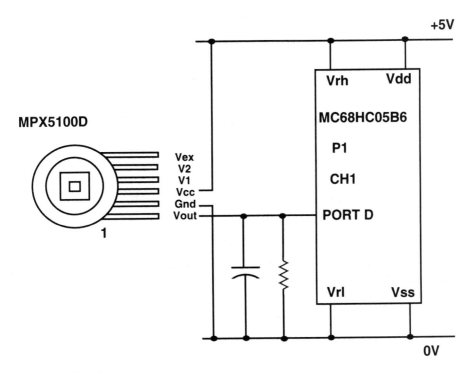

Figure 4.9 Continued.

Table 4.2 A/D Bits and Dynamic Range

A/D (# of Bits)	LSB Weight % of FS	LSB Voltage for 5-V F.S.	Dynamic Range (in dB)
4	6.25	300 mV	24.08
8	0.3906	19.5 mV	48.16
10	0.0977	4.90 mV	60.12
12	0.0244	1.20 mV	72.25
14	0.00610	305 μV	84.29
16	0.00153	75 μV	96.33

antialiasing filter attenuates frequencies above the Nyquist frequency and keeps the response below the noise floor.

Sigma-delta (or delta-sigma) converters are based on digital filtering techniques and can easily be integrated with DSP ICs. Sigma-delta (Σ-Δ) converters sample at a frequency much higher than the Nyquist frequency. Figure 4.10 shows the block diagram of an oversampled first-order sigma-delta ADC [23]. The analog input is summed at the input

Table 4.3 A/D Converter Architectures

ADC Type	Typ. # of Bits	Conversion Rate	Relative Die Area
Folding (flash)	8	100 MHz	14
Successive approximation (Hardware-driven)	12	300 kHz	10
Sigma-delta (Σ-Δ)	16	24 kHz	8
Successive approximation (Software-driven)	12	10 kHz	7
Dual-slope	12	3 kHz	3
Single-slope	12	3 kHz	1

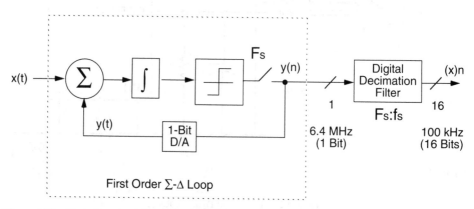

Figure 4.10 Block diagram of first order sigma-delta ADC.

node with the difference of the output of the 1-bit DAC. The resulting signal is provided to an integrator and then to a 1-bit quantizer (ADC). A second-order Σ-Δ consists of two integrators, two summers, and the 1-bit quantizer. Second- or third-order Σ-Δ modulators reduce the baseband noise level even further than a first-order Σ-Δ unit.

The output of the Σ-Δ converter is averaged by applying it to the input of a digital decimation filter. The digital decimation filter performs three functions. It removes out-of-band quantization noise, which is equivalent to increasing the effective resolution of the digital output. Secondly, it performs decimation, or sample rate reduction, bringing the sampling rate down to the Nyquist rate. This minimizes the amount of information for subsequent transmission, storage, or signal processing. Finally, it provides additional antialiasing rejection for the input signal.

4.4.2 ADC Performance

Frequently, the sensor requires high resolution at moderately fast conversion rates. In such cases, signal gain ranging and/or offset zeroing with 8-bit ADCs can be used in a cost-effective manner. Input ranging/zeroing allows for an increase in dynamic range without compromising the total conversion time or significantly increasing the silicon area, and hence cost. As shown in Figure 4.11, 10 bits of signal input can be achieved by range amplification and/or offset zeroing prior to an 8-bit A/D converter. Since the ranging/zeroing is typically defined by the settling time of the operational amplifier, the total conversion time remains small in comparison to a full 10-bit A/D conversion. Eight bits of accuracy is not a limitation depending on the application, as many processes normally operate around a rather small portion of the full-scale signal. A high degree of accuracy is required in the operating range, especially if the sensor is being used as a feedback element. By adjusting the input gain and offset, an 8-bit ADC can be placed at the optimum measurement point, but still be able to react to sudden excursions outside the normal range [21].

ADC resolution can also be increased with circuit techniques. For example, a circuit that uses both ADC and DAC channels of an MCU and a two-stage amplifier has been designed that increases the effective resolution of an 8-bit ADC to 12 bits for pressure measurements [24]. Standard components were used for the MCU and amplifier circuit. An applied pressure in the midrange of a 200-kPa sensor originally measured with an error of 0.3 kPa was subsequently reduced to 0.1k Pa using this approach.

Another consideration in choosing an ADC is specifying the requirements within the noise sources of the system. These limitations may be due to a less than optimum printed circuit layout, unstable power sources, nearby high-energy fields, and/or use of devices without good signal/noise and power supply CMRR. It is important to first assess the noise floor for the system and then compare this to the desired resolution of the signal. If the combined system's design and noise floor conditions create a minimum required signal that must be greater than 5–10 millivolts, only 8–9 bits of resolution can be determined within a 5V system. Using a higher resolution ADC will not provide any additional data unless

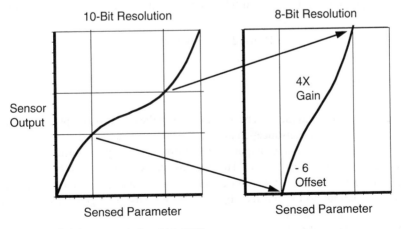

Figure 4.11   Increasing the accuracy of an 8-bit ADC.

the conversion speed of the ADC and processor throughput allows multiple sampling with data correction. Furthermore, this process must also meet the desired bandwidth, power consumption, and system response [25]. Oversampling and averaging is therefore desired because it is not possible to separate the noise from the signal using a single A/D sample/conversion.

4.4.3 A/D Accuracy/Error Implications

An amplified pressure signal supplied to an 8-bit ADC provides an example of the combined capability of the amplifier and ADC. The resulting A/D conversion is related to the pressure input by the following equation:

$$\text{count} = [V_{FS} - V_{Offset}] * 255/[V_{RH} - V_{RL}] \qquad (4.2)$$

Where $V_{FS} - V_{Offset}$ is the sensors full-scale span voltage, 255 is the maximum number of counts from the 8-bit converter, and $V_{RH} - V_{RL}$ is 5V based on using the same 5V supply as the MCU. Using the same reference voltage for the ADC and the sensor minimizes the number of additional components, but does sacrifice resolution. For those instances where greater resolution is required, separate lower voltage references should be provided for the ADC and/or the sensor.

For a sensor with a 0.25 to 4.75V output, the maximum number of counts available at the output register will be:

$$\text{count (full scale)} = 229$$

A full-scale pressure of 15 psi with 5.0V supply results in a system resolution of

$$15 \text{ psi}/229 = 0.066 \text{ psi/count}$$

4.5 SUMMARY

Several circuits and techniques can be used to amplify the low-level output that is inherently available from sensor or transducer elements. The circuits that were discussed are primarily used for interfacing the analog output of the sensor to the digital world of the microcontrollers and digital signal processors. The ongoing industry effort to improve performance and simplify the interface for sensors either as stand-alone units or integrated with the sensors will continue to expand the applications for sensors in general and simplify the design of future smart sensors.

Sleep-Mode, Rail-to-Rail, and ZIP·R trim are trademarks of Motorola, Inc.

REFERENCES

[1] Dunn, B., and R. Frank, "Guidelines for Choosing a Smart Power Technology," *PCI '88*, Munich, Germany, June 6–8, 1988.

[2] Klier, G., "Signal Conditioners: A Brief Outline," *Sensors*, Jan. 1990, pp. 44–48.

[3] Conner, D., "Monolithic Instrumentation Amplifiers," *EDN*, March 14, 1991, pp. 82–88.

[4] Williams, J., "Good Bridge-Circuit Design Satisfies Gain and Balance Criteria," *EDN*, Oct. 25, 1990, pp. 161–174.

[5] Schultz, W., "Amplifiers for Semiconductor Pressure Sensors," *Proc. of Sensors Expo West*, San Jose, CA, March 2–4, 1993, pp. 291–298.

[6] Motorola Application Note AN1324, *Pressure Sensor Device Data*, DL200/D, Rev. 1, 1994.

[7] Motorola Data Sheet MC33102/D, 1991.

[8] Motorola Data Sheet MC33201/D, 1993.

[9] Swager, A. W., "Evolving ADCs Demand More from Amplifiers," *Electronic Design News*, Sept. 29, 1994, pp. 53–62.

[10] Baher, H., "Microelectronic Switched Capacitor Filters," *IEEE Circuits and Devices*, Jan. 1991, pp. 33–36.

[11] Dunn, W., and R. Frank, "Automotive Silicon Sensor Integration," *SAE SP-903 Sensors and Actuators 1992*, Detroit, MI, Feb. 24–28, 1992, pp. 1–6.

[12] Winkler, C., and J. Baum, "Barometric Pressure Measurement Using Semiconductor Pressure Sensors," *Proc. of Sensors Expo West*, San Jose, CA, March 2–4, 1993, pp. 271–283.

[13] Motorola Application Note AN1318, *Pressure Sensor Device Data*, DL200/D, Rev. 1, 1994.

[14] Cianci, S., "Airborne Optical Encoders," *EDN Products Edition*, Jan. 16, 1995, pp. 9–10.

[15] Kandler, M., J. Eichholz, Y. Manoli, and W. Mokwa, "Smart CMOS Pressure Sensor," *Proc. of the 22nd International Symposium on Automotive Technology & Automation #90185*, Florence, Italy, May 14–18, 1990, pp. 445–449.

[16] Iida, M., Y. Isobe, Y. Yoshino, and I. Yokomori, "Electrical Adjustments for Intelligent Sensor," *SAE 900491, International Congress and Exposition*, Detroit, MI, Feb. 26–March 2, 1990.

[17] Senturia, S. D., "Microsensors vs. ICs: A Study in Contrasts," *IEEE Circuits and Devices*, Nov., 1990, pp. 20–27.

[18] Frank, R., and J. Staller, "The Merging of Micromachining and Microelectronics," *Proc. of 3rd International Forum on ASIC and Transducer Technology*, Banff, Alberta, Canada, May 20–23, 1990, pp. 53–60.

[19] Staller, J. S., and W. S. Cumberledge, "An Integrated On-chip Pressure Sensor for Accurate Control Applications," *IEEE Solid-State Sensors and Actuators Workshop*, Hilton Head, SC, 1986.

[20] Frank, R., "Two-Chip Approach to Smart Sensors," *Proc. of Sensors Expo*, Chicago, IL, 1990, pp. 104C1–104C8.

[21] Frank, R., J. Jandu, and M. Shaw, "An Update on Advanced Semiconductor Technologies for Integrated Smart Sensors," *Proc. of Sensors Expo West*, Anaheim, CA, Feb. 8–10, 1994, pp. 249–259.

[22] Hoeschele, Jr., D. F., *Analog-to-Digital and Digital-to-Analog Conversion Techniques*, New York, NY: John Wiley & Sons, 1994.

[23] Park, S., "Principles of Sigma-Delta Modulation for Analog-to-Digital Converters," Motorola APR8/D, Rev. 1, 1993.

[24] Motorola Application Note AN1100, *Pressure Sensor Device Data*, DL200/D, Rev. 1, 1994.

[25] Johnson, R. N., "Signal Conditioning for Digital Systems," *Proc. of Sensors Expo*, Philadelphia, PA, Oct. 26–28, 1993, pp. 53–62.

Chapter 5

Using MCUs/DSPs to Increase Sensor IQ

"We could use cow chips instead of microchips and save millions."
—Dilbert™ by Scott Adams

5.1 INTRODUCTION

Several semiconductor technologies are available to improve the accuracy and quality of the measurements and to add diagnostics and other intelligence to any type of sensor. Foremost among these technologies are: microcontrollers (MCUs), digital signal processors (DSPs), and application-specific integrated circuits (ASICs). Dedicated sensor signal processors are usually an adaptation of one of these approaches. These technologies also have the potential to allow for a fully integrated (monolithic) smart sensor. Before taking this next rather large step, it is important to understand the technologies that are available, their contribution to smart sensors, and their ability to provide a higher level of intelligence (and value) to sensors.

5.1.1 Other IC Technologies

ASIC technology utilizes computer-aided design (CAD) software tools to achieve custom circuit designs. This technology consists of programmable logic devices (PLDs) for low circuit density (< 5,000 gates), gate arrays for medium density (< 100,000 gates), and standard cells for high-end custom circuits [1]. ASIC devices combine high density and integration of full custom designs with relatively low cost and fast design turnaround. A custom system on a chip utilizing "core" microprocessor cells combined with analog, memory, and additional logic functions can address specific sensing requirements such as fluid-level sensing.

Mixed-signal ASICs combine analog with digital capability. For example, Customer Defined Arrays (CDA™) is a design methodology developed to achieve higher integration and performance levels by utilizing mixed technologies and architectures on the same chip.

Bipolar, BiCMOS, and CMOS technology with megacell, standard cell, and gate arrays can be combined to achieve maximum design flexibility. The combination can provide product features such as 50k gate density, 4-layer metal, 80%+ silicon utilization, 90 picosecond gate delays, 400 signal pads, and up to 256k bits of high-speed memory.

5.1.2 Logic Requirements

The shift of the logic requirements from a centralized computer to nodes in decentralized systems is creating the need for smart sensors. Sensor-driven process control systems that eliminate human operators and increase the precision of the process will play an important role in the manufacturing of the semiconductors that control them. A new architecture and interface have been proposed by researchers at the University of Michigan to address this application [2]. Figure 5.1 shows the block diagram of the key elements of this proposal. Amplification and A/D conversion were discussed in Chapter 4. The communications interface will be covered in Chapter 7. This chapter will use existing MCU and DSP

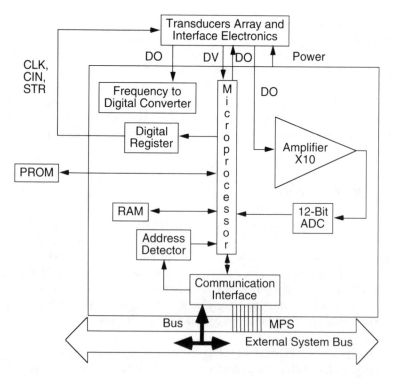

Figure 5.1 Block diagram of monolithic sensor interface design. (*After:* [2].)

products to demonstrate the remaining elements. Other system components that can be obtained with an MCU or DSP will also be discussed.

5.2 MCU CONTROL

Single-chip microcontrollers (MCUs) combine microprocessor (MPU) computing capability, various forms of memory, a clock oscillator, and input/output (I/O) capability on a single chip, as shown in Figure 5.2 [3]. Microcontrollers provide flexibility and quick time-to-market for numerous embedded control systems and for smarter sensing solutions. The programmability and wide variety of peripheral options available in the microcontroller provide design options. These options can offset the cost of the additional component by eliminating other components or providing features that would otherwise require far more components. In addition, these high-volume products enable systems to achieve low cost, high quality, and excellent reliability.

5.3 MCUs FOR SENSOR INTERFACE

In addition to the basic features of a microcontroller, a number of custom modules are integrated on the same chip to increase the utilization of the process, reduce printed circuit board space, and increase the functionality for a specific application. MCU features that

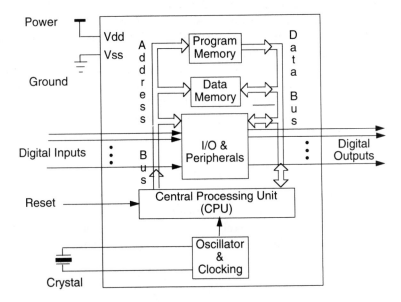

Figure 5.2 Basic microcontroller block diagram.

have a significant impact on sensor system performance will be discussed in this section. These include analog input capabilities, A/D conversion techniques, processing bandwidth, electronically programmable trim, onboard memory, power conservation, and improved electromagnetic compatibility (EMC) and control of radio frequency interference (RFI). For purposes of discussion, the MC68HC05 family of 8-bit microcontrollers will be used to explain the various MCU features. The same features, possibly with different specifications, perform similarly on other families of MCUs and higher performance 16-bit and 32-bit products as well.

5.3.1 Peripherals

Existing MCU technology contains a variety of peripherals or hardware options that enhance the capability of the MCU. Peripherals enable the MCU to obtain information from sensors and control output devices. Some of the most common peripherals are general purpose I/O ports, timers, and serial ports.

Timers usually measure time relative to the internal clock on the chip or an externally provided clock signal. The clock on the chip is controlled by an on-chip oscillator that operates at up to 4.0 MHz at 5V or 1 MHz at 3V. A more complex timer can generate one or more pulse-width modulated (PWM) signals, measure the pulse width, and generate additional output pulse trains.

Two basic serial ports are the serial communications interface (SCI) and the serial peripheral interface (SPI). The SCI is a simple 2-pin interface that operates asynchronously. Data is transmitted on one pin and received on the other. Start and stop bits synchronize communications between two devices. The SCI port is a universal asynchronous receiver transmitter (UART) that can be used with an RS-232 level translator to communicate with personal or other types of computers over fairly long distances.

The SPI port requires a third pin to provide the synchronizing signal between the control chip and an external peripheral. This type of communication is usually on the same board. Standard SPI peripherals are available from many manufacturers and include A/D converters, display drivers, electrically erasable programmable read only memory (EEPROM), and shift registers.

5.3.2 Memory

Various types of memory can be integrated on a chip, including RAM, ROM, EPROM, EEPROM, and flash memory. Semiconductor memory is based on a single transistor or cell that is on or off to generate a bit that is either a one or a zero. Memory is classified as either volatile or nonvolatile. Volatile memory is not stored when the power is disconnected to the MCU. Nonvolatile memory is stored when power is disconnected. The amount of memory in a chip is usually rated in kilobytes (1 KB = 1,024 bits). Increasing the amount of memory increases the chip size and the chip cost. Some types of memory, such as EEPROM, can significantly increase the process complexity and also add to the cost.

Random access memory (RAM) can be read or written (changed) by the CPU and is volatile [3]. Read only memory (ROM) can be read but not changed. This nonvolatile memory is included in the design (masked layout) of the chip. Reprogramming a chip once it has been designed is a common practice to correct errors in the original software, to upgrade to improve system performance, or to adjust for variation that could have occurred since the system was initially installed. Erasable programmable ROM (EPROM) can be changed by erasing the contents with an ultraviolet light and then reprogramming new values. This nonvolatile memory has a limited number of erasure and reprogramming operations. One-time-programmable ROM (OTP) is the same as EPROM except that it is packaged in a lower cost opaque package. Since ultraviolet light cannot penetrate the package, this memory cannot be erased after it is programmed. Electrically erasable programmable ROM (EEPROM) is a nonvolatile memory that can be changed by using electrical signals. Typically, an EEPROM location can be erased and reprogrammed thousands of times before it wears out. One of the newest types of memory is flash memory. Flash memory is nonvolatile memory that is easily reprogrammed in the system faster than EEPROM.

5.3.3 I/O

I/O (input/output) is a special type of memory that senses or changes based on external digital elements and not the CPU [3]. I/O ports connect these external elements to the CPU and provide control capability for the system. I/O can be either parallel (transferring eight bits at a time to the MCU) or serial (transferring data one bit at a time).

General purpose I/O connections (pins) can either be used as an input or an output. A number of pins are typically grouped together and called a port. The function of each pin is determined by the program. Program instructions evaluate the logic state of each input and drive outputs to logic one or zero to implement the control strategy. Input-capture and output-compare functions in the MCU simplify the design of the control strategy.

Input-capture is used to record the time that an event occurred. By recording the time for successive edges on an input signal, software can determine the period and/or the pulse width of the signal. Two successive edges of the same polarity are captured to measure a period. Two alternate polarity edges are captured to measure a pulse width [4].

Output-compare is used to program an action at a specified time. For example, an output is generated when the output-compare register matches the value of a 16-bit counter. Specific duration pulses and time-delay are easy to implement with this function.

5.3.4 Onboard A/D Conversion

Various types of ADCs were discussed in Chapter 4. An ADC is frequently integrated with the MCU. For example, the typical ADC in the HC05 family of MCUs consists of an 8-bit successive approximation converter and an input-channel multiplexer. Some of the channels are available for input and the others may be dedicated to internal test functions. For example, the 8-bit ADC of the MC68HC05B6 has a resolution of about 0.39%. The reference supply for the converter uses dedicated input pins to avoid voltage drops that would occur by loading the power supply lines and subsequently degrading the accuracy of the A/D conversions. To achieve ratiometric conversion, the +5V supply to the sensor is also connected to the V_{RH} reference input pin of the ADC, and the ground is referenced to V_{RL}. There is an 8-bit status control register, which is used to indicate that the oscillator and current sources have stabilized and that the conversion has been completed. The results of the conversion are stored in a dedicated 8-bit register.

For those instances where higher resolution is required, the SPI port of the 68HC05 allows external circuitry to be interfaced. For example, an integrated circuit such as Linear Technologies' LTC1290 connected to the SPI clock, data in, data out, and one additional programmable output pin provides a 4-wire interface for a 12-bit data conversion. The data are transferred in two 8-bit shifts to the 68HC05 in 40 µs. By adding the 12-bit capability, the resolution is improved from 0.39% to 0.0244%.

The successive approximation register (SAR) is the most popular method of performing A/D conversions due to its fast conversion speed and ease of use with multiplexed input signals. The 8-bit SAR ADC on the MC68HC05P8 MCU has the timing shown in Figure 5.3 and can be driven from the processor bus clock or an internal RC oscillator running at approximately 1.5 MHz [5]. This operating frequency makes the overall time to access and convert one signal source approximately 16–32 µs. Using Nyquist criteria for 2X sampling per cycle, such an A/D rate could be used for input sources up to approximately 31 kHz, which is well above that needed for many pressure, temperature, and acceleration sensors. With this increased bandwidth, additional samples can be taken and averaged to reduce effects of random noise sources and aliasing from higher frequency components. A simple averaging of four consecutive samples with this ADC can take less than 128 µs for an effective bandwidth of approximately 3.9 kHz. Since this ADC also has a hardware-driven SAR, each conversion can be started and performed without decreasing

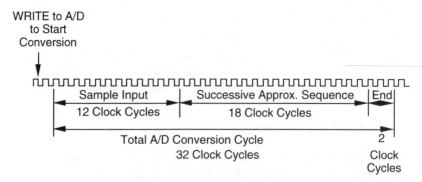

Figure 5.3   A/D process timing considerations for MC68HC05P8.

CPU processing power. Meanwhile, the CPU can process prior A/D samples until it is interrupted with a new conversion result. Some variations of this ADC on the higher performance 68HC11 family have hardware capable of automatically storing four consecutive readings into four separate results registers.

Another A/D consideration is the use of a sample and hold of the signal before the SAR operation begins. The ADC in the HC05 family performs this function in (typically) 6 to 12 μs. If the input is sampled too long, the signal must not change more than 1/2 LSB (least significant bit) during sampling in order to limit errors. This maximum change of 1/2 LSB in the input signal during the sample time can severely limit input bandwidth (considerably more than the minimum 2X sampling Nyquist criteria).

A further advantage of an ADC design with an internal RC oscillator is the ability to optimize power versus bandwidth. This is accomplished by providing a short A/D conversion cycle while running the main processor at a much lower bus frequency. This power-saving approach combined with others can increase useful battery life in portable applications.

5.3.5 Power-Saving Capability

Another advantage of MCU hardware and software is a variety of power-saving approaches. Varying the processing speed or stopping processing altogether can have a significant impact on overall power consumption. In addition, the ability to operate at lower voltages also reduces the power consumption. Reducing the power consumption involves more than reducing the supply current while running processor code.

The 68HC05 MCU family has two additional modes of operation to reduce power consumption, called the WAIT and STOP modes. In the WAIT mode, the processor bus is halted but the on-chip oscillator and internal timers are left in operation. The device will return to normal operation following any interrupt or an external reset. The WAIT mode in the MC68HC05P8 device reduces current consumption by nearly a factor of two at any given processor speed. In the STOP mode, both the processor and the on-chip oscillator are halted. This device will only return to normal operation following an external interrupt or reset. The STOP mode in the MC68HC05P8 reduces supply current consumption to less than 180 microamperes with a 5V dc supply operating at temperatures below 85°C.

An example demonstrates the importance of modes of operation and ADC performance versus power consumption of the signal processor. Consider a signal-processing example using a procedure that requires 100 CPU instructions following an A/D conversion of a pressure sensor with a 500-Hz bandwidth. The MC68HC05P8 modes of operation allow several approaches, as shown in Figure 5.4 [6]. Assuming a typical CPU cycle of 1 μs (1-MHz bus), the MC68HC05P8 will require 32 μs for an A/D conversion and an additional 300 μs for the digital procedure (approximately 3 cycles per instruction). A total system process time is 332 μs and would produce a 6X sampling of a 500 Hz signal.

If the MC68HC05P8 in this example is operated at a 1-MHz bus speed at 5V dc over

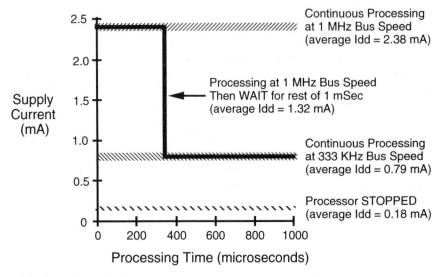

Figure 5.4 Processing mode effects on power consumption.

the temperature range of -40°C to +85°C, continuous conversions and processing will typically consume an average of 2.38 mA of supply current. However, if the device is only allowed to sample, convert, and process at the 2X Nyquist criteria of once every millisecond, it could sample and process in 332 μs and then go into the WAIT mode. The on-chip timer would wake it up each millisecond. The resulting supply current would drop to an average of about 1.32 mA, as shown in Figure 5.4. Another alternative to save power would slow the system clock to process a sample every millisecond. In this case, the CPU clock could be dropped to 3 μs (333-kHz bus) while the supply current to the MC68HC05P8 would drop to about 0.79 mA, as shown in Figure 5.4. Other combinations of bus speed and WAIT times will require average supply currents from 0.79 to 1.32 mA.

The STOP mode is useful in cases where the oscillator's restart time of (typically) 4064 CPU cycles is not a significant portion of the sampling/process time and data acquisition and processing is requested by an external source. For example, this occurs if a serial signal is transmitted every second following a temperature reading. Using the STOP mode, the supply current for the previous example would drop from about 2.38 mA for continuous 1-MHz bus operation to an average current near 0.18 mA. The sample, convert, process, and send sequence occurs at 1 MHz, then stops and waits for another external request to restart the sequence almost a second later.

5.3.6 Local Voltage or Current Regulation

Onboard voltage or current regulation is important to sensors that are not ratiometric since the variation in supply voltage over a -40°C to 125°C operating range can be greater than

±5%. The availability of analog control circuitry with 40V standoff capability allows the integration of a 5V series pass regulator on the MCU (for example, the MC68HC705V8) [6]. Availability of a similar shunt regulator allows a two-wire, self-protected and self-powered system to be designed using only a sensor and an MCU. At higher levels of integration, such analog voltage and/or current regulation can both reduce component count and improve accuracy. Furthermore, these regulation schemes can be dynamically altered to improve functionality or reduce power consumption.

5.3.7 Modular MCU Design

A methodology has been developed that allows custom microcontrollers to be designed to specifically address the requirements of a particular application [6]. The customer-specified integrated circuit (CSIC) approach differs from an application-specific IC in the performance and density that can be achieved. Typical ASIC chip solutions utilize (1) a family of basic elements designed to handle a variety of requirements and (2) automated design tools to reduce the design cycle time. Both of these attributes contribute to a larger die size. The automated chip assembly methods may also compromise analog performance when low-level signals are involved.

CSICs utilize standard functional subsystems that have a proven field history in applications including the demanding environment of automobiles. The chassis consists of pretested reusable blocks based on existing 68HC05 MCUs. These modular blocks include RAM, ROM, EPROM, and EEPROM from existing modules as well as serial communications modules; a variety of display drivers such as liquid crystal display (LCD), vacuum fluorescent display (VFD), or light emitting diodes (LED); timers; and A/D and D/A converters (see Figure 5.5). These modules can be modified, or existing modules from previously developed products can be utilized. Also, new modules can be designed to meet a specific customer requirement. The last case, "design-to-order," is only used for a small portion of the total design. The preferred approach utilizes hand-packed, highly optimized, and field-tested circuits to achieve short design cycle time (which averages about 6 months), first time success, and low end-product cost.

The CSIC approach minimizes or in some cases eliminates nonrecurring engineering costs and it allows a custom chip to be designed that can cost as low as $1. In many cases, one of the existing 150+ designs is acceptable to provide a "semicustom" solution without requiring a new design. The number of available MCU options and a methodology that allows quick customization of microcontrollers has been in place for several years to address meeting the specific requirements of various applications.

5.4 DSP CONTROL

Digital signal processors (DSPs) have hardware arithmetic capability that allows the real-time execution of feedback filter algorithms. In contrast, MCUs use look-up tables to

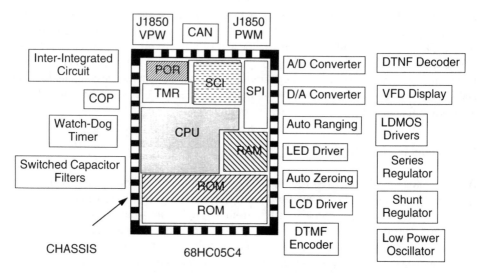

Figure 5.5 Custom MCU and various building blocks.

approximate filter algorithms with inherent limitations of flexibility and accuracy [1]. A typical DSP, like the DSP56000/1, has a 97-ns instruction. This capability allows a peak execution rate of 10 MIPS (millions of instructions per second), which is 5 to 10 times the performance of conventional MCUs. Additional features of the 56000/1 include:

- 144 dB dynamic range (336 dB possible on intermediate basis);
- 24-bit wide data paths;
- 62 instruction mnemonics;
- 3 internal communication peripherals;
- 18 interrupts.

The need for real-time processing in several systems is causing control system engineers to evaluate and use DSP technology for the control function. This growing class of functions cannot work effectively with traditional table look-up and interpolate functions to make the control decision. Instead, a multiply accumulate (MAC) unit allows state estimator functions to be implemented with an algorithm defining the state. However, as shown in Table 5.1, MCUs have an advantage over DSP units in many areas except real-time operating efficiency [7]. The question that system designers must answer is, "What sample rate and performance level are required for the application?" Also, the possibility of combining the new system with a previously developed or concurrently developed system should be evaluated. Depending on the nature of the system(s) and availability of control alternatives, a change from MCU to DSP may not be required.

DSP technology is evolving in two ways. First, DSP-like performance can be

Table 5.1 DSP Versus MCU Architecture Differences

Characteristic	DSP	MCU
Ease of programming		Better
Boolean logic		Better
Real-time operating efficiency	Better	
Code maintainability		Better
Peripherals availability		Better

approached when a time processor unit (TPU) is included in an MCU design. This is possible in higher performance MCUs. The TPU is a programmable microstate machine that addresses requirements for computation and greatly reduces the overhead on the main processor. This allows the main processor to calculate strategy-related items and not have to decide when a specific activity is initiated, such as firing the next sparkplug in an automotive engine control system. In some applications, the main signal processor will be a DSP. In other cases, multiple devices can be integrated on the same chip, especially with the capability of integrating an increasing number of transistors per chip. For example, multiple CPUs are used for fault tolerance in safety systems such as automotive antilock braking systems.

Second, in high-end applications such as an automotive near-obstacle detection system or noise cancellation system, a dedicated 24-bit DSP is proving to be a good solution. However, code portability is minimal in today's level of DSPs with the strategy closely linked to the I/O and the architecture. Also, generating code for a DSP, in general, is more difficult than generating code for an MCU. However, the MAC unit programmed in C requires less effort. As more sophisticated tools are available to deal with DSPs at the behavioral level, the current difficulties in programming fast Fourier transforms (FFT) and developing filters for DSPs will be simplified.

Figure 5.6 shows a 24-bit digital signal processor and its architecture. This device has program memory and two data memories. This DSP has the same kind of instruction set that is common throughout the Motorola microcontroller family. Using similar instruction sets makes it easier to write code and minimizes the discontinuity for programmers familiar with the MCU family. For example, the multiply, add, move, branch, and bit test are the same. However, the advanced features of the chip require new instructions.

The DSP56001 has been used to develop an intelligent sensor for checking the pitch of tapped holes in motor blocks. The sensor replaced a mainframe computer that was checking one hole per minute. By providing DSP power in each sensor, the inspection system increased the rate to 100 holes per minute [8]. The sensor was developed using a DSP design kit that is the size of a credit card. Voltage monitoring, battery backup circuitry, and a watchdog timer are features of the design kit and the DSP-based sensor.

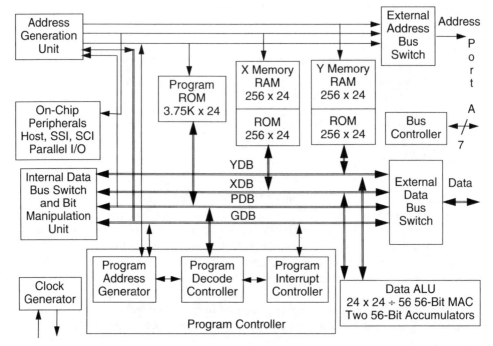

Figure 5.6 24-bit DSP architecture.

5.4.1 Algorithms versus Look-Up Tables

MCUs use look-up tables to store values that are accessed when the program is running. Algorithms are used to correct for variations from expected results and to implement a control strategy. The speed of accessing the information from a table or performing a calculation determine the response time of the sensor input to the MCU/DSP portion of the system. This can be the limiting factor to initiating a change to the output in an MCU/DSP controlled system.

A control + sensor combination can implement an electronically programmable trim as an alternative to laser trimming. However, all trimming and calibration processes for a sensor require some form of data conversion by the MCU or DSP. The time it takes to perform these conversions by mathematical calculations or get data from a look-up table must be within the control system's ability to respond to the sensed input. In the application, real-time trimming can be implemented to allow adaptive control at the sensor level. This will improve the accuracy of a sensor that has shifted after some time in operation.

5.5 TECHNIQUES AND SYSTEMS CONSIDERATIONS

Increased precision can be obtained for sensors by characterizing devices over temperature and pressure (or acceleration, force, etc.) and storing correction algorithms in MCU memory. The MCU can convert the measurement to display different units (i.e., psi, kPa, mmHg, or inches of water for pressure measurements). Other techniques use the MCU's capability to improve linearization, to provide PWM outputs for control, and to provide autozeroing/autoranging. The operating frequency and switching capability of the MCU must be considered in system design. Lastly, the MCU's computing capability can be used in place of sensor(s) when sufficient information exists. These system aspects will be explored in this section.

5.5.1 Linearization

Sensor nonlinearity can be improved by the use of table look-up algorithms. The variation in sensor signal caused by temperature can also be improved by using an integrated temperature sensor and a look-up table to compensate for temperature effects while linearizing the output, nulling offsets, and setting full-scale gain from information stored in an EEPROM. Look-up tables can be implemented in masked ROM, field-programmable EPROM, or onboard EEPROM.

Compensation for nonlinearity and the number of measurements during the test and calibration procedure can be simplified if a nonlinear output correlates with a sensor design parameter. For example, a strong correlation was found between the span and linearity of a pressure sensor with a thin diaphragm [9]. As Figure 5.7 shows, the nonlinearity increased to almost 5% with the highest span units.

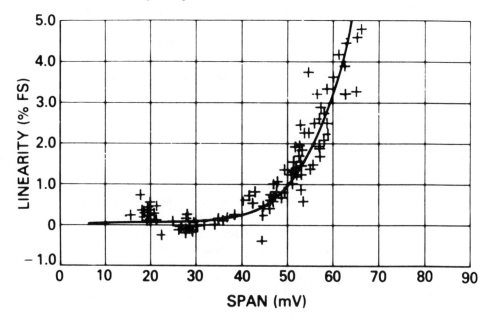

Figure 5.7 Span versus linearity for pressure sensor output.

Analytical techniques were investigated to improve the nonlinearity. A polynomial regression analysis was performed on 139 sensor samples ranging from 30 to 70 mV full-scale span to determine the coefficients B_0, B_1, and B_2 in the formula:

$$V_{out} = V_{off} + (B_0 + B_1 \cdot P + B_2 \cdot P^2 + B_3 \cdot P^3 + ...) \tag{5.1}$$

where B_0, B_1, B_2, and B_3 are sensitivity coefficients. The second-order terms were sufficient for calculations to agree with measured data with a worst case value for calculated regression coefficient = 0.99999. The relationship of these values to the span allowed a piecewise linearization technique with four windows to reduce most sensors to less than 0.5%. These calculations could be included in the MCU look-up table for improved accuracy in an application. Others have also investigated linearization in great detail as a general means to improve sensor accuracy [10, 11].

5.5.2 PWM Control

The pulse width modulation (PWM) output from the MCU can be used to convert an analog sensor output to a digital format for signal transmission in remote sensing or noisy environments [12]. Figure 5.8 shows the simple, inexpensive circuitry used to create a duty cycle that is linear to the applied pressure. The MCU-generated pulse train is applied to a ramp generator. The frequency and duration of the pulse can be accurately controlled in software. The MCU requires input-capture and output-compare timer channels. The output-capture pin is programmed to output the pulse train that drives the ramp generator while the input-capture pin detects edge transitions to measure the PWM output pulse width. The pulse width changes from 50 to almost 650 μs for zero to full-scale output for this sensor.

5.5.3 Autozero and Autorange

Combining a sensor and an MCU to perform a measurement that otherwise would be less accurate or more costly than other available alternatives is feasible today. The cost of many MCUs is comparable or lower than the micromachined sensors that provide their input signal. For example, a signal-conditioned pressure sensor has been combined with a 68HC05 MCU to measure 1.5 inch or less of water with an accuracy of 1% of the full-scale reading [13]. The MCU provides software calibration, software temperature compensation, and dynamic-zero capability. Also, a digital output compatible with the SPI protocol is provided for the pressure measurement.

Autoreferencing can be performed by the MCU to correct for common-mode errors, especially with low-level signals. Autoreferencing uses the MCU logic and clock signal combined with a digital-to-analog converter (DAC) and counter. A signal from the MCU initiates the counter. The DAC provides a sample and hold and programmable voltage

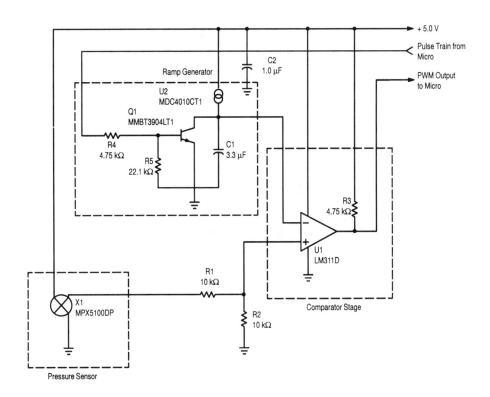

Figure 5.8 PWM output pressure sensor schematic.

source. The sensor output is summed with the autoreference correction at the input of an amplifier to obtain a corrected output to the system.

A calibration-free method for pressure sensors has been designed using a calibrated pressure sensor, an MCU with an integral ADC, and two additional ICs [14]. As shown in Figure 5.9, two input channels and one output port are used in the calibration portion of this system. All errors from the instrument amplifier are canceled in this circuit. For a measurement that only requires a full-scale accuracy of ±2.5%, the offset of the pressure sensor can be neglected and the system does not require any calibration procedure. For a ±1% measurement, the full-scale output is set at 25°C.

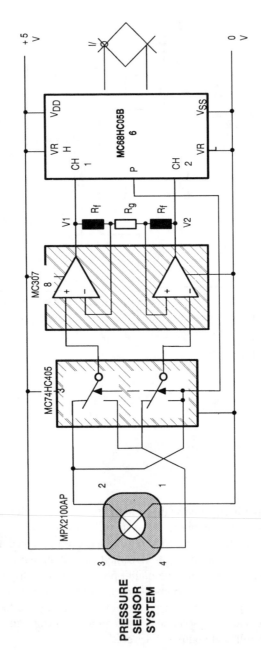

Figure 5.9 Calibration-free pressure sensor system.

5.5.4 Diagnostics

One of the more valuable contributions that the MCU can make to the sensor's functionality is the ability to self-test, analyze status, diagnose, and report problems to an operator or other systems that use the sensor's output. The problem could be as simple as a warning for maintenance or calibration or an indication of catastrophic failure that would not allow the system to function when required. For example, an automotive air bag system may operate for years without requiring deployment. However, knowing that the sensor is capable of providing the signal to indicate a crash event is part of the safety the system can routinely provide the driver and passenger. In the air bag system, a dash indicator lamp is activated by the MCU to indicate that it has performed a system-ready analysis.

5.5.5 EMC/RFI Reduction

Radiated radio frequency interference (RFI) and electromagnetic interference (EMI), both the transmitting and receiving of unwanted signals or electromagnetic compatibility (EMC), of a device is becoming an important design consideration with increasingly higher levels of system integration and higher processing bandwidths. These problems can be reduced at the component level with smaller radiation loops and a smaller number of signal lines. Power management in the MCU can also eliminate some self-induced effects that generate RFI to the sensor element. An example of power management is "quiet-time sampling." This circuit technique performs analog switching when digital switching is not present [6]. Cleaner samples of data are taken by halting high current, high-frequency activity while the sensor input is being measured. Normal digital processing functions continue once these values are in memory.

5.5.6 Indirect (Computed not Sensed) versus Direct Sensing

Inputs to the MCU can be manipulated to provide additional data for a system. For example, the MCU can use an input pressure signal to provide maximum pressure, minimum pressure, an integrated (averaged) pressure, and time-differentiated pressure data to a system. An accelerometer signal processed by a signal processor can be integrated once to provide velocity and twice to provide displacement information. Also, a single sensor input to an MCU can replace several switches sensing the same parameter and provide programmable switch points for outputs.

The ability to compute rather than sense is among the solutions that MCUs bring to control applications. For example, in three-phase motor control systems a Hall effect sensor is used to sense the location of the magnetic field for each phase or the rotor speed in induction motors. An MCU uses this signal to switch output drivers for PWM control. The sensors in this system have been a target for cost reduction for many years. Recent solutions have eliminated the sensors by using the MCU to compute the rotor speed and slip angular frequency from other available information, including the primary resistance [15].

5.6 SOFTWARE, TOOLS, AND SUPPORT

Creating a new approach or alternative to existing control technologies requires much more than the architecture. The software that is used to program the control portion and tools that allow the system to be developed are equally important. Portable code (software compatibility) is essential if future end products may require migration to a higher performance MCU. Also, keeping the sensor's design and process simple, and separate from the MCU, allows the implementation with other available processors.

One of the main advantages of using existing MCUs or DSPs is the development tools that already exist and allow the designer to quickly and easily develop both system hardware and software. For example, data entry, program debugging, and programming of an MCU's OTP, EPROM, and EEPROM can be accomplished by utilizing an evaluation module (EVM) (refer to Figure 5.10) interfaced to a host computer such as an IBM-compatible PC or Apple Macintosh® [16]. All of the essential MCU timing and I/O circuitry already exist on the EVM to initiate the development process. MCU code may be generated using a resident one-line assembler/disassembler or may be downloaded to the user program RAM through the host or terminal port connectors. The code can be executed by several debugging commands in the monitor.

A development unit that allows emulation of several 68HC05 versions is also available. The CDS8/05 development system uses a PC host as an intelligent node. This unit can be powered by an automobile cigarette lighter for in-vehicle development.

The availability of freeware through an electronic bulletin board assists in software development. This information, the development tools, and the documentation have been established over several years of customer usage. As a result, the capability exists to develop the smartest sensor that can be defined today.

5.7 SENSOR INTEGRATION

The fourth level in advancing integration (refer to Figure 1.7 in Chapter 1) is to migrate the developed MOS interface onto the MCU itself. The basic sensor element is interconnected in close proximity to this enhanced MCU within the same package. Now the sensor itself is the only element that does not need to be MOS-process-compatible. This fourth level is also the optimum point at which to consider including any output drive capabilities required by the specific sensing and control application onto the MCU. This type of power integration could have been accomplished in the earlier Level III except that standard production MCUs with external drivers could have been more effectively used at the lower levels. At Level IV, there is a definite transition from standard product MCUs into customization for a specific application or market. Therefore, all aspects of the total system

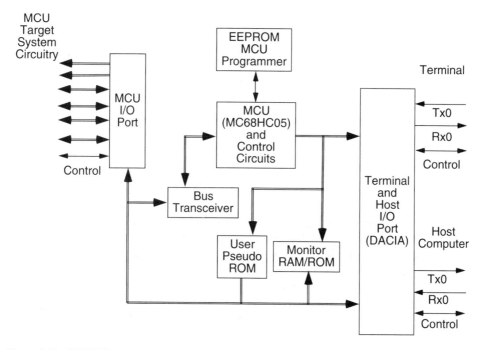

Figure 5.10   MC68HC05EVM development tool for 68HC05 MCU.

should be considered for integration at this level in order to maximize the benefits of the added design and development investment.

Table 5.2 shows the implications of different process defects on the net cost of a combined solution [6]. Present industry MOS fabrication processes have typically between 0.6 to 1.0 defects/cm^2 with the best-in-class at 0.2 defects/cm^2. To demonstrate the costs of combining a sensor with an MCU, consider a small-size (100 mils by 100 mils) MCU processed at 0.8 defects/cm^2 and a 100 mils by 100 mils sensor processed at 4.0 defects/cm^2. The example MCU will have a relative cost of 1.05 versus a 100 mils by 100 mils device processed with zero defects. The sensor will have a relative cost of about 1.29. This gives a combined two-chip relative cost of 2.34. If the sensor and MCU are combined on a single die using a sensor process that creates 4.0 defects/cm^2, the resulting die would be about 128 mils by 128 mils and have a relative cost of 2.50. This cost increase of only 1.07 times the combined die cost may be less significant than the added assembly and test costs of two separate devices.

However, the cost tradeoff changes dramatically as the MCU die size increases. If the same 100 mils by 100 mils sensor is combined with a larger 300 mils by 300 mils MCU with a relative cost of 15.15, the relative cost of the combined two-chip set rises to 16.44. Combining the two chips into an equivalent die size of 311 mils by 311 mils and processing

Table 5.2 Theoretical Relative Die Cost for Sensor Integration

	Best in Class*	Typical MOS Industry*	Expected Level of Sensor Process Development	
Defects per cm^2	0.2	0.8	2.0	4.0
Die size (mils)				
100 by 100	1.01	1.05	1.14	1.29
128 by 128	1.70	1.81	2.05	2.50
300 by 300	10.85	15.15	27.69	65.11
311 by 311	11.31	16.17	30.89	75.28

*Theoretical, 6-inch wafers. Die cost of 1.0 for 100 mils by 100 mils die and zero defects/cm^2.

at the higher defectivity sensor process will yield a relative cost of 75.28, or 4.58 times higher than the two-chip solution. Integrating a (typically) higher defectivity sensor process onto another chip becomes increasingly more costly as the total die size increases. In all cases, adding a sensor to another silicon device will add some cost due to the added silicon area for a fixed batch size (wafer diameter). However, the cost penalty of adding a sensor on a chip decreases if the defectivity of the sensor process decreases or the wafer size increases. As these improvements occur, the added cost can be traded off against packaging and assembly costs for the two-chip approach.

An interim approach to the fully integrated sensor on an MCU is shown in Figure 5.11. The sensor interface for the accelerometer is initially a stand-alone circuit that uses the same design rules and process as the MCU. This provides a shorter design cycle and lower initial cost for the sensor. Once the end product (containing the combination of this sensor and the MCU) has achieved market acceptance and a minimum run rate, a phase II design (Level III in Figure 1.7) is possible. The sensor interface is treated as a predesigned building block for a fully integrated sensor signal processor using the modular design approach discussed earlier in this chapter.

5.8 SUMMARY

Trends in the development of microelectronics for MCUs, DSPs, and ASICs have been for faster, more complex signal processing and reduced feature size (critical dimensions)—into the submicron range. These goals have led to lower supply voltages (3.3V and less). In addition, customer requirements for increased and more easily programmed memory and reduced development cost and design cycle time are affecting the design and methodology used to design these products. The resulting improvements can be useful in the design of smart sensors and can be cost-effective by using other capabilities provided by the MCU, DSP, or ASIC. Communicating the data from smart sensors is among the capabilities that can be integrated on chip and is such a critical part that Chapter 7 will be dedicated to this aspect of the control logic.

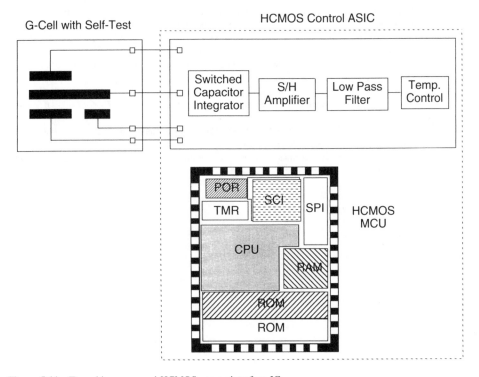

Figure 5.11 Two-chip sensor and HCMOS sensor interface IC.

REFERENCES

[1] Frank, R., and J. Staller, "The Merging of Micromachining and Microelectronics," *Proc. of 3rd International Forum on ASIC and Transducer Technology*, Banff, Alberta, Canada, May 20–23, 1990, pp. 53–60.

[2] Najafi, N., and K. Wise, "An Organization for Interface for Sensor-Driven Semiconductor Process Control Systems," *IEEE Trans. on Semiconductor Manufacturing*, Vol. 3, No. 4, Nov. 1990, pp. 230–238.

[3] Sibigtroth, J. M., *Understanding Small Microcontrollers*, Motorola Technical Bulletin M68HC05TB/D, Rev. 1, 1992.

[4] *M68HC05 Microcontroller Applications Guide*, Motorola M68HC05AG/AD.

[5] Benson, M. et al., "Advanced Semiconductor Technologies for Integrated Smart Sensors," *Proc. of Sensors Expo '93*, Philadelphia, PA, Oct. 26–28, 1993, pp. 133–143.

[6] Frank, R., J. Jandu, and M. Shaw, "An Update on Advanced Semiconductor Technologies for Integrated Smart Sensors," *Proc. of Sensors Expo West*, Anaheim, CA, Feb. 8–10, 1994, pp. 249–259.

[7] Krohn, N., R. Frank, and C. Smith, "Automotive MPU Architectures: Advances and Discontinuities," *Proc. of the 1994 Congress on Transportation Electronics*, SAE P-283, Oct. 1994, Dearborn MI, pp. 71–78.

[8] "MiniKit 56001 Provides DSP Power to Low-Volume Products," *DSP News*, Vol. 6, Issue 3, 3Q93, Motorola, 1993, p. 5.

[9] Derrington, C., "Compensating for Nonlinearity in the MPX10 Series Pressure Transducer," AN935 in *Pressure Sensor Device Data*, Motorola DL200/D, Rev. 1, 1994.

[10] Hille, P., R. Hohler, and H. Strack, "A Linearisation and Compensation Method for Integrated Sensors," *Sensors and Actuators*, A 44, Netherlands: Elsevier Sequoia, 1994, pp. 95–102.

[11] Heintz F., and E. Zabler, "Application Possibilities and Future Chances of "Smart" Sensors in the Motor Vehicle," 890304 in SAE *Sensors and Actuators* SP-771, SAE, Warrendale, PA, 1989.

[12] Jacobsen, E., and J. Baum, "Using a Pulse Width Modulated Output with Semiconductor Pressure Sensors," Motorola Application Note AN1518, 1994.

[13] Ajluni, C., "Pressure Sensors Strive to Stay on Top," *Electronic Design*, Oct. 3, 1994, pp. 67–74.

[14] Burri, M., "Calibration Free Pressure Sensor System," Motorola *Pressure Sensor Device Data*, DL200/D, Rev. 1, 1994.

[15] Kanmachi, T., and I. Takahashi, "Sensor-Less Speed Control of an Induction Motor," *IEEE Industry Applications Magazine*, Jan./Feb. 1995, pp. 22–27.

[16] Frank, R., "Two-Chip Approach to Smart Sensors," *Proceedings of Sensors Expo*, Chicago, IL, 1990, pp. 104C-1 to 104C-8.

Chapter 6

Communications for Smart Sensors

"Behold the people is one, and they all have one language; and this they begin to do: and now nothing will be restrained from them, which they have imagined to do."

—Genesis 11: 6

6.1 INTRODUCTION

The increasing interest in smart sensors is a direct result of the need to communicate sensor information in distributed control systems. Unfortunately, numerous protocols have been defined for data communication in control systems. Industry standards have been and are being developed for various applications. Within a given market segment, several proposals are vying for acceptance. A few of these protocols have already been implemented in silicon hardware. This chapter will provide background information, identify many of the proposed protocols, and describe silicon chips that have been developed to support system designs.

6.2 BACKGROUND AND DEFINITIONS

Data communication from sensor inputs must be analyzed recognizing that the sensor is a part of an entire control system. Considerable effort has been expended by several industry committees and individual companies to generate acceptable protocols that will support their requirements for distributed control. System functionality and the ability to use available silicon hardware, software, and development tools must be considered when selecting a protocol for a given application.

Several new terms and definitions must be used to describe the communication of system information. Protocol is an agreed-upon set of rules for communications. The bus connects internal or external circuit components and can be serial or parallel, but the serial approach is more common. Multiplexing (MUX) is the combining of several messages for transmission over the same signal path. Access to the bus is obtained through an arbitration process. Bit-by-bit arbitration is also called collision detection. Contention is the ability to gain access to the bus on a predetermined priority. Latency is guaranteed access (with maximum priority) within a defined time. A deterministic system can predict the future behavior of a signal. The data link controller (DLC) is the silicon implementation of the protocol that handles all of the communications requirements.

The system goal of interoperability will permit sensors or actuators from one supplier to be substituted for those from another manufacturer. The different types of networks are shown in Figure 6.1. The star, ring, and linear (tree, multidrop) are common topologies.

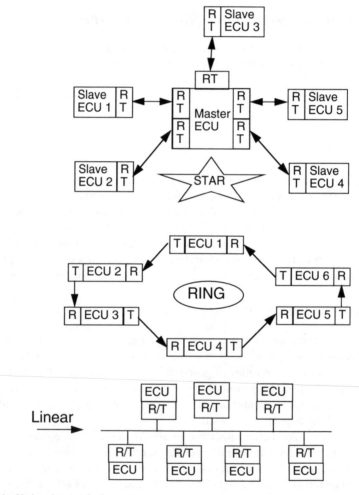

Figure 6.1 Various bus topologies.

The use of master-slave relationships, although an important part of control networks, is giving way to the use of more distributed intelligence in many applications.

The 4- to 20-mA standard of the Instrument Society of America has been the analog data transmission standard for over twenty years. However, digital techniques can provide more system functionality and increased noise immunity for signal transmission. Many digital interface formats already exist for point-to-point or multidrop communications, as shown in Table 6.1 [1]. These protocols have seen limited used in computer-controlled systems. The EIA-485 is most often used for digital field bus applications. It is the physical layer in Profibus, Topaz, and Bitbus. The EIA-485 is a balanced-line (differential) data transmission over a single twisted-wire pair [1]. However, examining the various market segments, an even wider variety of standards exists.

6.2.1 Background

The International Standards Organization (ISO) has defined the open systems interconnect (OSI) model that describes seven layers for data networking. The layers and their functions are shown in Figure 6.2 [2]. Available and proposed standards typically use all or a portion of these layers to define the standard.

In a distributed system, a node contains the sensor (or actuator), the hardware for local

Table 6.1 Point-to-Point and Multidrop Standards

Parameter	EIA-232-C	EIA-432-A	EIA-422-A	EIA-485
Mode of operation	Single-ended	Single-ended	Differential	Differential
Number of drivers	1	1	1	32
Number of receivers	1	10	10	32
Maximum cable length (ft)	50	4,000	4,000	4,000
Max data rate (bps)	20k	100k	10M	10M
Max common-mode voltage	±25V	±6V	6V	12V
			-0.25V	-7V
Driver output	±5V min	±3.6V min	±2V min	±1.5V min
	±15V max	±6V max		
Driver load	±3 kΩ to 7 kΩ	450 Ω min	100 Ω min	60Ω min
Driver slew rate	30 V/μs max	Externally controlled	NA	NA
Driver output	500 mA to Vcc	150 mA	150 mA	150 mA to GND
Short-circuit current limit	or GND	to GND	to GND	250 mA to -8V or 12V
Receiver output resistance (on)	NA	NA	NA	120 kΩ
(high Z state) (off)	300Ω	60Ω	60Ω	120Ω
Receiver input resistance	3 kΩ to 7 kΩ	4 kΩ	4 kΩ	12 kΩ
Receiver sensitivity	±3 V	±200 mV	±200 mV	±200 mV

After: [1].

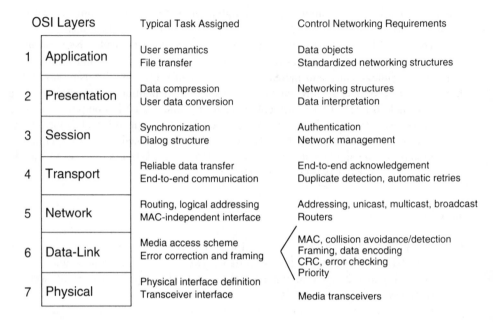

OSI Layers		Typical Task Assigned	Control Networking Requirements
1	Application	User semantics File transfer	Data objects Standardized networking structures
2	Presentation	Data compression User data conversion	Networking structures Data interpretation
3	Session	Synchronization Dialog structure	Authentication Network management
4	Transport	Reliable data transfer End-to-end communication	End-to-end acknowledgement Duplicate detection, automatic retries
5	Network	Routing, logical addressing MAC-independent interface	Addressing, unicast, multicast, broadcast Routers
6	Data-Link	Media access scheme Error correction and framing	MAC, collision avoidance/detection Framing, data encoding CRC, error checking Priority
7	Physical	Physical interface definition Transceiver interface	Media transceivers

MAC = media access control
CRC = cyclic redundancy check

Figure 6.2 ISO's open system interconnect (OSI) model.

computations, and the network interface [3]. The term *sensor bus* is frequently used when sensors are connected in a system using a multiplexed bus offering data-link-level multiplexing and allowing packets from different senders to be sent to different receivers. A higher level *sensor network* uses all layers of the OSI model to provide more information and simplify the user's system design and maintenance activity.

6.3 SOURCES (ORGANIZATIONS) AND STANDARDS

Several universities have developed standards for communication. For example, the University of Michigan has proposed the Michigan parallel standard (MPS) [4]. Researchers at Delft University of Technology have developed a serial communications protocol [5]. The Integrated Smart-Sensor (IS^2) bus is a mixed analog/digital two-line serial bus interface. However, a protocol requires more than definition and demonstration hardware. Accepted usage by a number of manufacturers ultimately determines the real standards, which may in fact be de facto standards. University-developed protocols may achieve acceptance with the adaptation of industrial users, but this process takes longer than utilizing existing industry-sponsored standards. The focus of this chapter will be standards that are already supported by industry. A list of the more common standards, including those developed by universities, manufacturers, and standards organizations is shown in Table 6.2.

6.4 AUTOMOTIVE PROTOCOLS

Due to its large volumes, the automotive segment has driven specifications resulting in the actual implementation of protocols in several products. In automobiles, information from one sensor and/or data from one system can be communicated with other systems using multiplex wiring to reduce the number of sensors and the amount of wire that is used in a vehicle. Two predominant protocols have emerged as standards, but several other protocols exist for specific manufacturers' applications, as shown in Table 6.1. The Society of Automotive Engineers (SAE) has established SAE J1850 as the standard for multiplexing and data communications in the U.S. automobiles. However, trucks and OBDII (On-Board Diagnostics II) are based on the controller area network (CAN) protocol developed by Robert Bosch GmbH.

The SAE Vehicle Network for Multiplexing and Data Communications (Multiplex) Committee has defined three classes of vehicle networks: Class A, Class B, and Class C [6]. Class A is for low-speed applications such as body lighting. Class B is for data transfer between nodes to eliminate redundant sensors and other system elements. Class C is for high-speed communications and data rates typically associated with real-time control systems. The data rates and latencies for these three classes are summarized in Table 6.3.

6.4.1 SAE J1850

SAE J1850 was approved as the standard protocol for U.S automakers in 1994. J1850 defines the application, data link, and physical layers of the OSI model. J1850 specifies two implementations: a pulse-width modulation (PWM) and variable pulse width (VPW) version. The differences are indicated in Table 6.4. SAE J1850 is supported by a number of other SAE specifications that are referenced in J1850, such as the message strategy document, J2178.

In addition to the SAE bit-encoding techniques specified in Table 6.4, other common techniques include 10-bit NRZ (nonreturn to zero), Bit-stuf NRZ, L-Man (Manchester), E-Man, and modified frequency modulation (MFM). These techniques differ based on variable synchronizing, arbitration, transitions per bit, maximum data rate, oscillator tolerance, and integrity. SAE J1850 network access is nondestructively prioritized by bit-by-bit arbitration for either protocol option [6]. A frame is defined as one complete

Table 6.2 Protocols and Sponsors in Various Market Segments

Automotive	Sponsor
J-1850	Society of Automotive Engineers
J-1939 (CAN)	Society of Automotive Engineers
J1567 C^{2}D	SAE (Chrysler)
J2058 CSC SAE	(Chrysler)
J2106 Token Slot	SAE (GM)
CAN	Robert Bosch GmbH
VAN	ISO
A-Bus	Volkswagen AG
D^{2}B	Philips
MI-Bus	Motorola

Industrial	Sponsor
Hart	Rosemount
DeviceNET	Allen-Bradley
Smart Distributed Systems	Honeywell
SP50 Fieldbus	ISP + World FIP = Fieldbus Foundation
SP50	IEC/ISA
LonTalk	Echelon Corp.
Profibus	German DIN
ASI bus	ASI Association
InterBus-S	InterBus-S Club
Seriplex	Automated Process Control (API, Inc.)
SERCOS	VDW (the German machine tool manufacturers association)
IPCA	Pitney Bowes, Inc.

Building/Office Automation	Sponsor
BACnet	Building Automation Industry
LonTalk	Echelon Corp.
IBIbus	Intelligent Building Institute
Batibus	Merlin Gerin (France)
Elbus	Germany

Home Automation	Sponsor
Smart House	Smart House L.P.
CEBus	EIA
LonTalk	Echelon Corp.

Protocol University	Sponsor
Michigan Parallel Standard	University of Michigan
Integrated Smart-Sensor Bus	Delft University of Technology
Time-Triggered Protocol	University Wien, Austria (Automotive)

Table 6.3 Automotive Network Classes

Class	Type	Data Rate	Latency
A	Low	1 k to 10 Kbps	20–50 ms
B	Medium	10 k to 100 Kbps	5–50 ms
C	High	10 k to 1 Mbps	1–5 ms

Table 6.4 Protocol Options in SAE J1850

Feature	1 & 3 Byte Headers	1 & 3 Byte Headers
Bit encoding	PWM	VPW
Bus medium	Dual wire	Single wire
Data rate	41.7 Kbps	10.4 Kbps
Data integrity	CRC*	CRC

* CRC= cyclic redundancy check

transmission of information. Within the frame, the header contains information regarding the message priority, message source, target address, message type, and in-frame response. This will be explained further in the next section. For SAE J1850, each frame contains only one message and the maximum length for a frame is 101 bit times. A power reduction or sleep mode occurs at a node if the bus is idle for more than 500 ms. Wake-up occurs with any activity on the bus.

6.4.2 CAN Protocol

CAN is a serial communications protocol developed by Robert Bosch GmbH that was originally designed for automotive multiplex wiring systems, especially high-speed data communications. CAN supports distributed real-time control with a high level of security and message integrity. It has also become attractive for use in lower speed and other distributed control applications.

The original CAN specification was announced in the 1980s. A revision, CAN 2.0, was announced in 1991. CAN 2.0 consists of an A and B part. Part A is known as CAN 2.0 A, CAN 1.2, and BasicCAN. Part A specifies an 11-bit identifier field, includes no specification for message filtering, and has a layered architecture description based on Bosch's internal model. Part B enhancements include an extended 29-bit identifier field, some message filtering requirements, and layer description based on the ISO/OSI reference model. The 29-bit identifier field allows the automotive standard protocol, SAE J1850, 3-byte headers to be mapped into the CAN identifier field. However, minimum CAN compliance is established by conformance to CAN 2.0 A only [7].

CAN is a multimaster protocol that allows any network node to communicate with any other node on the same network. Any node can initiate a transmission once it has determined that the network is idle. CAN properties include user-defined message prioritization. CAN is actually a nondeterministic system, but a guaranteed maximum latency for highest prioritization can be calculated. Lower priorities are determined on a statistical basis. CAN utilizes carrier sense, multiple access/collision resolution (CSMA/CR) for nondestructive collision resolution. The arbitration technique is bitwise and results in the highest priority message being transmitted with low latency time. The flexible system configuration allows user options that have led to CAN-developed systems in automotive and industrial applications [7].

The message format for CAN is a fixed-format frame with a variable number of data bytes. Zero to 8 data bytes are permissible. The minimum data frame length is 44 bit times. Four message types are defined: data frame, remote frame, error frame, and overload frame. The messages are routed to receiving nodes through the use of message identifiers and message filtering. Functional addressing allows multiple nodes to act on a single message. However, nothing in the hardware prevents the user from using physical addressing to achieve node-to-node addressing. Part of the error detection scheme is the acceptance of every message by all nodes or no nodes. Even if a component on the network is not concerned with the message that is transmitted, it must still receive the message, check the CRC, and acknowledge (ACK) acceptance. The data frame shown in Figure 6.3 is the most widely used frame. Priority is established in the identifier field based on the user's messaging strategy [7].

Data is requested through the use of a remote frame and transmitted through the data frame. The remote frame contains no data field and the remote transmission request (RTR) bit is sent recessive. Otherwise, the remote frame is identical to the data frame [8].

Error detection and error signaling are possible because CAN has a variety of built-in error counters that help contain errors and prevent nodes that have failures from restricting communications on the bus. A faulty transmitting node always increases its error counter more than other nodes. Therefore, the faulty node becomes the first to go "bus off." Automatic retransmission of corrupted messages minimizes the possibility of an undetected message. CAN also has the ability to distinguish between temporary errors and permanent node failures, which make it ideal for high-noise environments that occur in automotive and industrial applications [7]. The error message frame is shown in Figure 6.4. Bit error, stuff error, CRC error, form error, and acknowledge error can be detected. The error frame contains the superposition of error flags sent by various nodes on the network and the error delimiter. Both passive and active error codes are possible in CAN, depending on the status of the node [8].

The CAN protocol does not address the physical layer requirements. The transceiver is not specified, but rather is left to the user's network characteristics to define the requirements. (Bit timing rules in the CAN specification must be met). Accepted physical media include, but are not limited to: twisted pair (shielded or unshielded), single wire, fiber-optic cable, or a transformer coupled to power lines. RF transmitters are also being

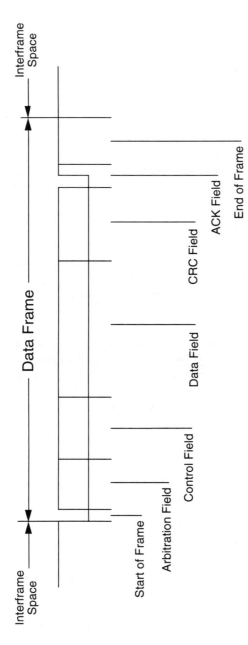

Figure 6.3 CAN data frame.

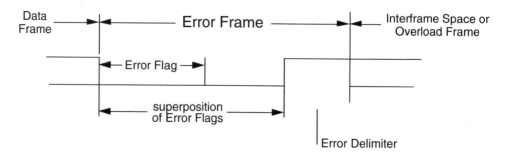

Figure 6.4 CAN error frame.

developed by some users for CAN systems. Acceptable transmission rates range from 5 Kbps to 1 Mbps. The most widely used implementation to date is a twisted-pair bus with NRZ bit formatting. Twisted-pair CAN drivers are available that simplify system design [7].

6.4.3 Other Automotive Protocols

As shown in Table 6.1, other automotive protocols have been developed. Some of these are covered by SAE specifications. The implementation of these alternatives requires both a volume user that obtains benefit by using the alternate protocol, as well as a semiconductor manufacturer willing to implement the protocol in silicon hardware. For comparison, the A-bus and VAN will be discussed.

Automotive bit-serial universal interface system (A-bus) was developed by Volks-wagen AG [6]. Error detection is by bit only. Both single wire and fiber optics can be used for the transmission media. The maximum bus length is not specified, but it is typically 30m. The maximum date rate of 500 Kbps is considerably higher than SAE J1850.

Vehicle area network (VAN), developed by French automaker Renault, is being considered as an ISO standard. The transmission medium is twisted pair. The maximum data rate is user definable. A maximum of 16 nodes is possible with a maximum bus length of 20m. Bit encoding is L-MAN and E-MAN. The Manchester coding technique encodes a 1 with a high level for the first half of the bit time, and a 0 is encoded with a low level for the second half of the bit time. Consequently, a maximum of two transitions per bit time is possible.

6.5 INDUSTRIAL NETWORKS

The power of a distributed multiplex system is easily demonstrated in factory automation. In many cases, users have found that an installation that previously took a crew of electrical technicians several days to wire could be completed by one or two people in a day with a multiplex system. Also, these installations work successfully the first time, resulting in a considerable cost reduction. Wiring consists of a twisted pair of wires, power, and ground, which greatly simplifies the interconnection process. Once a system strategy is developed, nodes can be added or moved easily without reengineering the system. These nodes can include sensors as well as valves, motors, and lighting loads. The key is an open standard and plug-and-play capability [7].

The industrial market has even more proposed and developing standards than automotive. Fieldbus is the term for a nonproprietary digital two-way communications standard in the process automation industry. The Fieldbus specification will define the application, data link, and physical layers of the ISO model with some layer 4 services defined. A typical industrial application with Fieldbus as the highest performance level and a sensor bus at the lowest level is shown in Figure 6.5 [9]. However, Fieldbus has not been completed and semiconductor products are not available to implement the control nodes. Two protocols that are attracting a number of industrial users based on available silicon products are CAN and LonTalk™.

6.5.1 Industrial Usage of CAN

CAN has been adopted for use in industrial applications by manufacturers such as Allen-Bradley and Honeywell in their DeviceNET™ system and Smart Distributed System

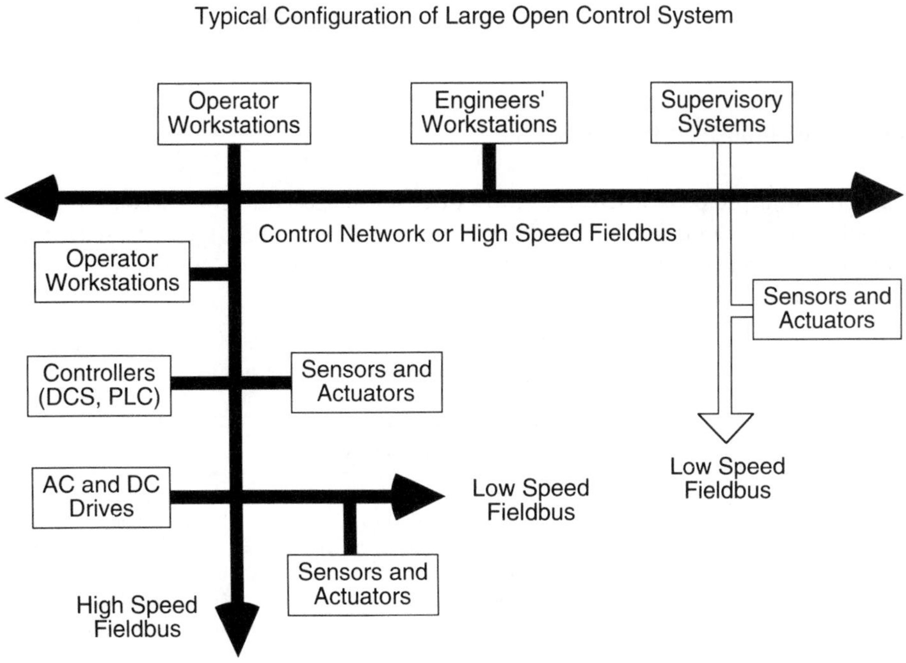

Typical Configuration of Large Open Control System

Figure 6.5 Fieldbus control system architecture. *(After:* [9].)

(SDS™), respectively. These communication networks have been developed as simpler, lower cost alternatives to Fieldbus, which is being developed as an industrial standard. However, Fieldbus is intended to handle a larger amount of data. All three networks are designed for real-time operation, but each handles a different amount of transmitted data. The different attributes of SDS, DeviceNet, and Fieldbus are shown in Table 6.5 [10].

The CAN protocol is used to implement both SDS and DeviceNet, but these two systems are not interoperable. The difference between them is in the physical layer. This layer is not defined in the CAN specification, allowing users to implement different approaches.

6.5.2 LonTalk™ Protocol

The LonTalk™ protocol developed by Echelon Corporation has received considerable support for several industrial and consumer applications. This protocol defines all seven layers of the OSI model. It uses differential Manchester coding with a data packet length of 256 bytes and can operate up to a maximum speed of 1.25 Mbps. The arbitration is predictive carrier sense multiple access (CSMA) and has collision detection and priority options. The LonTalk protocol can be used to support sensor network to Fieldbus requirements [11]. A LonWorks™ system works on a peer-to-peer basis and does not require a host CPU. Events occurring at a particular device are communicated directly between modules.

6.5.3 Other Industrial Protocols

Other industrial protocols listed in Table 6.1 offer advantages to distributed control systems. A brief summary of the most widely known, frequently mentioned, and better documented protocols will conclude this section. These include the Hart, Fieldbus, Profibus, Sercos, and Topaz protocols.

Table 6.5 Industrial Communications Network Comparison

Attribute	Sensor Bus	Device Network	Fieldbus
Network name	SDS	DeviceNet	WorldFIP, ISP
Target devices	Sensors	Pushbuttons, sensors, switches, drives, starters, valves	Smart transmitters, flowmeters, servovalves
Data packet	Typically 1 byte	Up to 8 bytes	Up to 1,000 bytes
Data rate	≤ CAN	Up to 500 Kbps	Up to 1–2.5 Mbps
Data quantity	Limited data	Moderate data	Large amounts of data
Interoperability	Vendor-specific	Open	Open

The highway addressable remote transducer (Hart) protocol was developed by Rosemount. It has a dedicated users group with several participating companies. The open multimaster protocol is compatible with the 4 to 20 mA current loop and also allows digital communication [12].

The recent combination of Interoperable Systems Project Foundation (ISPF) and WorldFIP North America in the Fieldbus Foundation has the potential to complete a Fieldbus specification that has been in development for over eight years. Fieldbus defines the user layer in the OSI model, making it more extensive than protocols that address only physical, data link, and/or application layers (i.e., CAN) [13].

Profibus (Process Field Bus) is a German DIN standard that supports up to 1.5-Mbps data rate. The deterministic system uses a master-slave hierarchy and has a capacity of 127 nodes. Arbitration is handled by command response. EIA-485 is used as the physical layer [12].

The SERCOS (serial, real-time communication system) network typically consists of eight nodes per ring. It operates at a speed of up to 1 Mbps and for a distance of up to 40m at this speed. It is a deterministic multimaster network with command response and broadcast for arbitration. The physical layer is fiber optic [12].

Topaz can handle up to 255 nodes and operate at a speed of 1 MHz. At 1 kilometer, the speed decreases to 500 Kbps. The multimaster system is deterministic and uses token passing for arbitration. The physical layer uses the EIA-485 interface [12].

6.6 OFFICE/BUILDING AUTOMATION

The BACnet protocol has been developed by the building automation industry. Ethernet, ARCNET, MS/TP and Lonworks are among the networks that could communicate on the proposed BACnet-compatible system being developed by the American Society of Heating, Refrigeration, and Air-Conditioning Engineers (ASHRAE). Building Energy Management Systems standards are also being developed. In addition, the Automatic Meter Reading Association (AMRA) is developing a standard for automatic meter reading. IBIbus has been developed by the Intelligent Building Institute [14].

Smart office buildings will offer a high degree of automation. Figure 6.6 shows the interconnection of several systems [14]. In the offices, nodes will sense changes in the environment and send status and control messages to other nodes in response to those changes. Power nodes will open or close dampers, change fan speed, and make other adjustments based on this information. Other aspects of these systems will be self-diagnostics, data logging, fire detection and sprinkler systems, energy usage monitoring, and security systems.

6.7 HOME AUTOMATION

The computer control of future homes is the goal of the Smart House project. Among the likely candidates for interfacing to the network are the heating, ventilation, and air con-

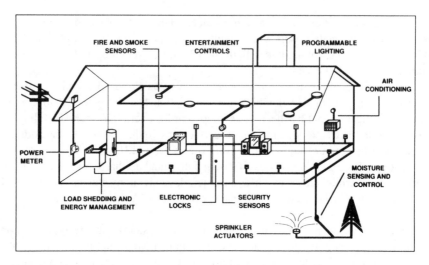

Figure 6.6 Building automation.

ditioning (HVAC) system; water heater; range; security; and lighting [15]. Remote meter reading and demand-side management by utility companies are among the driving forces for home applications. Protocol acceptance in this consumer environment is contingent on achieving low cost per node and ease of operation. The speed of these systems will range from low to high, depending on the devices that are connected to the system. Both the message size and message protocol complexity will be medium.

Smart house applications language (SHAL) has been developed that includes over 100 message types for specific functions. Dedicated multiconductor wiring is required in the home. The system can address 900 nodes and operates at a maximum of 9.6 Kbps [12]. Two additional contenders in this arena are CEBus and LonTalk.

6.7.1 CEBus

Consumer Electronics Bus (CEBus) was initiated by the Electronic Industries Association (EIA) Consumer Electronics Group. CEBus provides both data and control channels. It handles a maximum of 10 Kbps and has a growing acceptance in the utility industry [12].

6.7.2 LonTalk

LonTalk's acceptance in building automation, availability of full OSI layers, and interoperability are among the reasons that it is well-suited for the home automation environment.

Widespread acceptance will also drive the learning curve for cost reduction that this market requires.

6.8 PROTOCOLS IN SILICON

A few of the protocols that have been discussed have been implemented in silicon hardware and are available from multiple sources. In some cases, the protocol is a stand-alone integrated circuit. However, a protocol integrated into a microcontroller provides computing capability and a variety of system features that impact the cost-effectiveness of individual system nodes.

6.8.1 MCU with Integrated SAE J1850

A standard protocol provides sufficient volume potential for semiconductor manufacturers to design a variety of integrated circuits. In many cases, the highly integrated solution provides the lowest cost per node. For example, Motorola's MC68HC705V8, shown in Figure 6.7, integrates the J1850 communication protocol, on-chip voltage regulation, the physical layer interface, the J1850 digital aspect (VPW version), and other systems related functions. The integrated solution has many features that make system design easier and reduces both the component count and system cost.

The heart of the J1850 chip is Motorola's data link controller (MDLC) shown in Figure 6.8 [16]. The MDLC handles all of the communication duties, including complete message buffering, bus access, arbitration, and error detection [17]. Other J1850 versions have also been implemented by a number of suppliers [18].

6.8.2 MCU with Integrated CAN

The CAN protocol has been implemented on a variety of microcontrollers by several semiconductor manufacturers. The integrated CAN solution contains a variety of I/O, memory, and other system features. One key point to note about the CAN protocol is that although a set of basic features must be implemented in any CAN device, a number of extended features may also be implemented in silicon, depending upon the intended target application. The implications must be understood when considering interoperability and interchangeability. Several solutions have been designed and are offered by a number of semiconductor manufacturers. In fact, this broad product availability is among the reasons that industrial users are choosing CAN [7].

An example of the CAN protocol implemented in an MCU is provided by Motorola's M68HC05 microcontroller family. This CAN implementation conforms to the CAN 2.0 part A specification. All Motorola CAN (MCAN) devices include automatic bus arbitration, collision resolution, transmission retry, digital noise filtering, message I.D. filtering, frame

Figure 6.7 Systems integration in MC68HC705V8 IC (*courtesy of* Motorola, Inc.).

generation and checking, CRC generation and checking, as well as single transmission and multiple receive frame buffers [19]. As shown in Figure 6.9, the MCAN module contains full message transmit and receive buffers and limited message filtering capability [20]. The integrated bus interface includes input comparators and CMOS output drivers. However, an external transceiver may be necessary, depending on the user's application.

The CAN module is totally CPU-independent and may be integrated with a CPU core, other MCU hardware, and I/O functions into a single chip. Figure 6.10 shows the block diagram of a minimum version of a CAN module integrated in an MCU [21]. This integrated MCU version includes a wake-up feature on port inputs, a 16-bit timer with one input capture and one output compare, a computer operating properly (COP) watchdog, and is available in a cost-effective 28-pin SOIC package. Other features are available on this chip, and even more are included in the more complex versions shown in Table 6.6. Example

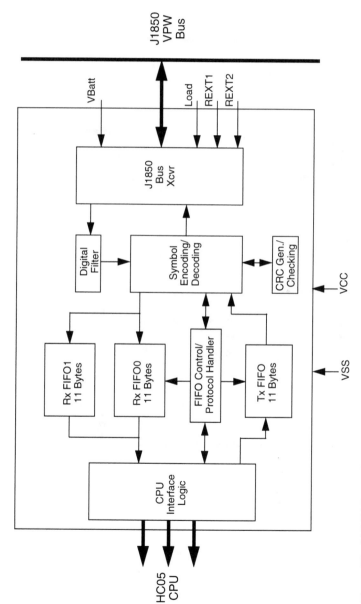

Figure 6.8 SAE J1850 data link controller.

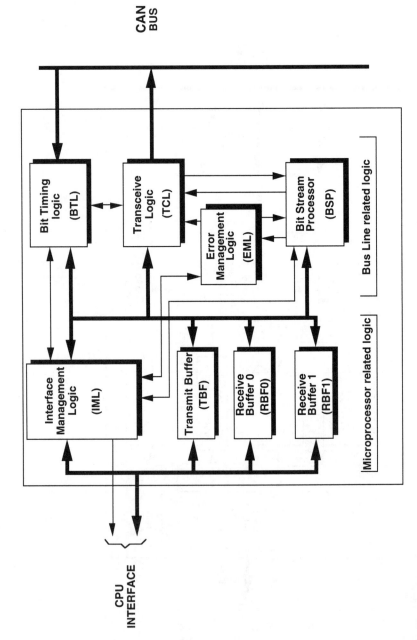

Figure 6.9 CAN module.

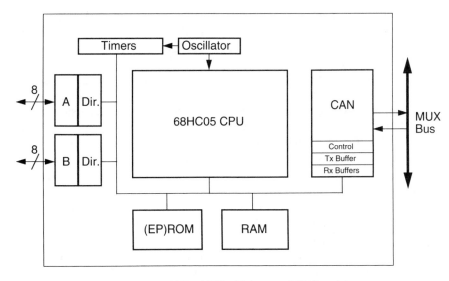

Figure 6.10   Block diagram of MC68HC05X4 MCU with integrated CAN module.

Table 6.6 Features of MCUs with Integrated CAN Modules

Feature	X4	X16	X32
MC68HC05 CPU	√	√	√
Basic CAN protocol (Rev. 2.0 part A)	√	√	√
8-bit ID mask register for optional message filtering	√	√	√
1 Tx message buffer	√	√	√
2 Rx message buffers	√	√	√
Bytes of user program (EP)ROM	4,112	15,168	31,248
Bytes of EEPROM	---	256	256
Bytes of use data RAM	176	352	352
Programmable input/output pins, ( ) input only	16	24 (8)	24 (8)
Port B wired OR interrupt operation	—	√	√
Serial communication interface (SCI)	—	√	√
8-channel ADC	—	√	√
16-bit timer with __ input capture + __ output compares	1 + 1	2 + 2	2 + 2
2 pulse-length modulation systems	---	√	√
COP watchdog (with COP enabled on reset option)	√	√	√
EPROM and EEPROM protection	---	√	√
Package (number of pins)	28	64/68	64/68

software drivers are available for these units, which provide an interface between application software residing in the MCU ROM (or EPROM) and the CAN module. These routines initialize the CAN module, transmit messages previously stored in RAM, and automatically handle received messages [7].

Memory, either in RAM, ROM, EPROM, or EEPROM, is required for proper CAN operation. The smaller -X4 version includes 4,096 bytes of EPROM (one-time programmable version) or ROM and 176 bytes of RAM. The -X16 and -X32 versions have 15,120 bytes of (EP)ROM, 256 bytes of EEPROM and 352 bytes of data RAM for more complex programs. The EEPROM on the -X16 also provides flexibility for addressing and programming over the network. A user can manufacture a number of identical modules and put the ID in EEPROM when each is installed to minimize inventory costs. The ADC allows easy interface for sensors and the pulse-width modulation (PWM) module can be used for digital speed control of power devices. The integrated timers can be used as part of the program control. Also, the buffering of the CAN module requires CPU intervention for each message received or transmitted. Due to the additional functionality designed into the MC68HC05X16, the CAN module occupies less than 20% of the active area of the chip, and this feature makes the integrated function very cost-effective [22].

6.8.3 Neuron Chips and LonTalk™ Protocol

The LonTalk protocol is an integral part of a Neuron™ IC, such as Motorola's MC143150 or MC143120. The Neuron IC is a communications and control processor with an embedded LonTalk protocol for multimedia networking environments in which received network inputs are used to control processor outputs. As shown in Figure 6.11, the MC143150 has three CPUs: one for the media access control (MAC) processor, one for the network processor, and one for the application processor [7].

The MAC processor handles layers one and two of the ISO protocol stack. This includes driving the communications subsystem hardware as well as executing the collision avoidance algorithm. This processor communicates with the network processor using network buffers located in shared memory [11].

The network processor implements layers three through six of the ISO stack. Network-variable processing, addressing, transaction processing, authentication, background diagnostics, software timers, network management, and routing functions are handled by this unit. In addition to communication with the MAC processor, it communicates with the application processor through application buffers.

The application processor executes the code written by the user and the operating services requested by user code. The primary programming language is Neuron C, a derivative of the C language. Enhancements in Neuron C for distributed control applications are supported in firmware. As a result, the user can write complex applications with less program memory.

Both Neuron chips include a direct-mode transceiver and 25 application I/O models.

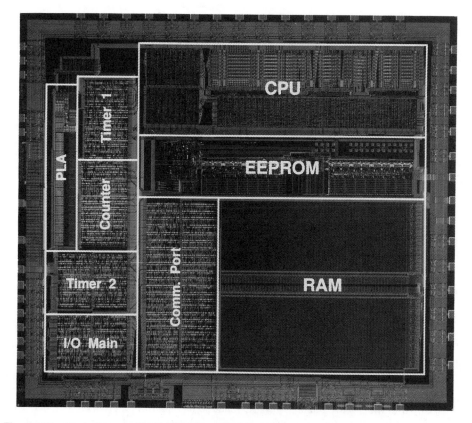

Figure 6.11   Block diagram of MC143150 Neuron™ chip (*courtesy of* Motorola, Inc.).

Other integrated functions include the network communication port, several I/O functions including two timers/counters, clocking and control capability, and (for the MC143150) 2 KB of RAM and 512 bytes of EEPROM. Fuzzy logic kernels have been written for the Neuron chip that can be used with support tools to easily implement fuzzy logic as part of the control portion of the distributed network [23].

6.8.4 MI-Bus

Motorola has developed a protocol called the MI-Bus for a low-cost network solution. This approach for class A operation is built around Motorola's HC05 and HC11 microcontrollers and is designed to achieve low cost in certain master-slave control applications. The MI-Bus protocol has been designed for direct control of (low frequency) power loads [24]. The master-slave arrangement allows one master to control as many as eight slaves. The MI-Bus

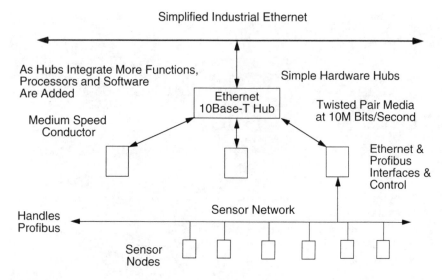

Figure 6.12 Ethernet and Profibus interface.

can be driven out of the I/O port. It should be noted that even with the considerable effort that has been expended on J1850 and CAN specifications and hardware implementations that have been developed, a need for an alternate solution can be identified by a customer with sufficient volume to justify the development of a custom solution.

The MCU initiates the protocol by issuing a sequence of 8 data bits, which contain the 3-bit address and 5-bit control. The three address bits allow up to eight devices to be addressed. After transmitting data, the MCU reads serial data from the selected device. Bus access method, frame sequence protocol, message validation, error detection, and default values are all defined for the MI-Bus.

6.8.5 Other MCUs and Protocols

The Profibus protocol is available in RAM microcode in a number of semiconductor devices. One single-chip local area network (LAN) solution has been developed to interface Ethernet and a variety of different sensor networks including Profibus [25]. As shown in Figure 6.12, network nodes can be connected to an Ethernet hub using twisted-pair cables. The MC68360 in this application has a 32-bit CPU that provides 4.5 MIPS at 25 MHz for power and flexibility in any existing Ethernet system. Since Ethernet has the leading number of installed LANs, interoperability is essential.

6.9 OTHER ASPECTS OF NETWORK COMMUNICATIONS

In addition to the protocols that have already been discussed, microcontroller manufacturers use serial communications protocols between ICs that are on the same printed circuit board.

These interfaces may also be considered for subsystems in communications networks. In using one of these MCU protocols or any of the previously discussed protocols, the need to transition between various protocols can occur, especially in complex systems. MCU protocols, the transition between protocols, and modular protocol solutions will be the final topics on communication in this chapter.

6.9.1 MCU Protocols

The serial peripheral interface (SPI) is a high-speed protocol that is used for synchronously transferring data between master and slave units in an onboard serial network. As shown in Figure 6.13, the SPI module is a defined building block that can be used for transmitting digital information in a system through a software-defined protocol. This module is already contained in a number of microcontroller families. The QSPI, or queued SPI, is an intelligent, synchronous serial interface with a 16-entry, full-duplex queue. It can continuously scan up to 16 independent peripherals and maintain a queue of the most recently acquired information without CPU intervention.

The serial communications interface (SCI) transmits using only two pins on MCUs with an SCI peripheral. The SCI protocol was used to develop the MI-Bus discussed in Section 6.8.4. This option will be discussed in more detail in the next section.

The inter-IC, or I^2C, is a two-wire half-duplex serial interface with data transmitted/received most significant bit first. The two wires are a serial data line (SDA) and a serial clock line (SCL). The protocol consists of a start condition, slave address, n bytes of data, and a stop condition. Each byte is followed by an acknowledge bit. The I^2C peripheral can be interfaced by means of a synchronous serial I/O Port or SIOP.

6.9.2 Transition between Protocols

Because several communication protocols exist, the transition between protocols is frequently required. A gateway node provides the transition between networks that use different protocols. In some cases, custom hardware and software gateways convert information from one system's language to that of the other system. Ideally, the MCU in this

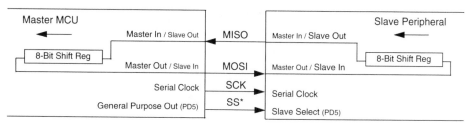

Figure 6.13 The serial peripheral interface (SPI).

node has one or both protocols, which allows an easy transition. In the worst case, the MCU uses a separate peripheral for each protocol. Figure 6.14 shows the gateway master from the CAN to an MI-Bus. On the master side, the SCI protocol common to many MCUs can be used. On the other side of the gateway, the MCU has a CAN module for communication on the CAN network. This allows control information to pass from the CAN network across to drive the output loads. On the slave side, the I/O controller provides the interface to the load. A stepper motor controller, the MC33192, was the first device to control loads in an automotive environment using the MI-Bus [24]. The integrated device has both the MI-Bus controller and a dual full-bridge driver for the stepper motor. Other modules can easily be designed to control other loads by simply revising the output portion of the MI-Bus device.

6.9.3 The Protocol as a Module

The protocol can be easily reconfigured into a new MCU design using modular design methodology described in Chapter 5. In addition to the CPU and other modules, serial communications protocols, including CAN, both versions of SAE J1850, SPI, SCI, and I^{2}C can be treated as modular peripherals. This allows new features to be easily added to an MCU with an integrated protocol as they are required for different applications. Also, a protocol can be easily added to an existing MCU design that will become part of a communication network.

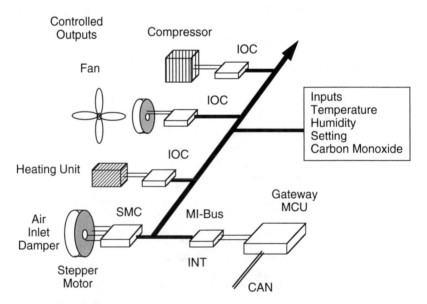

Figure 6.14 CAN to MI-Bus gateway and various sensors and loads.

6.10 SUMMARY

This chapter has discussed communication protocols for sensors. The ability for sensors to communicate with other portions of the control system will allow more intelligence at the sensor node and more distributed control. Protocols that have available silicon solutions have a higher likelihood of acceptance. Many types of standards have been developed; however, implementation options may limit their compatibility. An entirely new protocol developed for sensor-specific applications would have to be well-defined and accepted by sensor users to justify the development cost necessary to implement the protocol in silicon.

LonTalk and LonWorks are trademarks, and Neuron is a registered trademark of Echelon Corporation. DeviceNET is a trademark of Allen-Bradley Company, Inc. SDS is a trademark of Honeywell, Inc. All other trademarks are the property of their respective owners.

REFERENCES

[1] Pullen, D., "Smart Sensors: Adding Intelligence to Transducers," *Proc. of Sensors Expo West*, San Jose, CA, March 2–4, 1993, pp. 117–136.

[2] Raji, R. S., "Smart Networks for Control," *IEEE Spectrum*, June 1994, pp. 49–55.

[3] Madan, P., "LonWorks™ Technology for Interfacing and Networking Sensors," *Proc. of Sensors Expo*, Philadelphia, PA, Oct. 26–28, 1993, pp. 225–242.

[4] Najafi, N., and K. D. Wise, "An Organization and Interface for Sensor-Driven Semiconductor Process Control Systems," *IEEE Trans. on Semiconductor Manufacturing*, No. 4, Nov. 1990, pp. 230–238.

[5] Bredius, M., F. R. Riedijk, and J. H. Huijsing, "The Integrated Smart Sensor (IS²) Bus," *Proc. of Sensors Expo*, Philadelphia, PA, Oct. 26–28, 1993, pp. 243–247.

[6] Meisterfeld, F., "Chapter 26 - Multiplex Wiring Systems," *Automotive Electronics Handbook*, R. K. Jurgen, editor, New York, NY: McGraw-Hill, 1994, pp. 26.1–26.63.

[7] Frank, R., and C. Powers, "Microcontrollers with An Integrated CAN Protocol," *Proc. of Power Conversion And Motion*, Sept. 17–22, 1994, pp. 1–10.

[8] Terry, K., "Software Driver Routines for the Motorola MC68HC05 CAN Module," Motorola AN464, 1993.

[9] Chatha, A., "Fieldbus: The Foundation for Field Control Systems," *Control Engineering*, May 1994, pp. 77–80.

[10] Jost, K., "The Three-Day Car and Manufacturing Control Systems," *Automotive Engineering*, Nov. 1994, pp. 13–16.

[11] Neuron Chips, Motorola Brochure BR1134/D, Rev. 1, 9/93.

[12] Tonn, J., Internal Motorola Document, 1994.

[13] McMillan, A., "Fieldbus Interfaces for Smart Sensors," *Proceedings of First IEEE/NIST Smart Sensor Interface Standard Workshop*, Gaithersburg, MD, March 31–April 1, 1994, pp. 133–148.

[14] Sack, T., "Building Buses: The Plot Thickens," *Electrical Review*, Vol. 224, No. 20, Oct. 1991, pp. 28–29.

[15] Jancsurak, J., "Electronics Breakthroughs Now and On the Horizon," *Appliance Manufacturer*, May 1993, pp. 29–30.

[16] MC68705V8 Motorola Specification, Rev. 2, 1993.

[17] Powers, C., "Example Software Routines for the Message Data Link Controller Module on the MC68HC705V8," Motorola AN1224, 1993.

[18] Powers, C., "J1850 Multiplex Bus Communications Using the MC68HC705C8 and the SC371016 J1850 Communications Interface (JCI)," Motorola AN1212, 1992.

[19] *Protocols in Silicon*, Motorola Brochure FLDR86S/D, 10/92.

[20] CAN Module, Motorola Product Preview, 1990.

[21] MC68HC05X4/MC68HC705X4 Advanced Information, Motorola, 1993.

[22] Neumann, K.-T., ''The Application of Multiplex Technology,'' *Automotive Technology International '95*, pp. 199–205.

[23] ''Fuzzy Logic and the Neuron Chip,'' Motorola AN1225, 1993.

[24] Burri, M., and P. Renard., ''The MI-BUS and Product Family for Multiplexing Systems,'' Motorola EB409, 1992.

[25] O'Dell, R., ''Ethernet Joins the Factory Workforce,'' *Machine Design*, July 11, 1994, pp. 78–82.

Chapter 7
Control Techniques

"A machine is intelligent when it does the things of a common three year old."
—Marvin Minsky, creator of the concept of artificial intelligence

7.1 INTRODUCTION

Distributed control systems (DCS) have brought decision making to the sensor. The built-in intelligence for the smart sensor can take advantage of a variety of existing and developing control techniques. The extent to which new computing paradigms will affect sensors depends on the need for control at the sensor and the cost-effectiveness of using these new techniques. Existing control techniques have grown from programmable logic controllers and PC-based instruments. Proportional integral derivative (PID) control and the state machine are the basis of many control systems. However, the newest artificial intelligence approaches of fuzzy logic and neural networks will increasingly affect sensor-based systems. Fuzzy logic is already finding several applications in appliances and automobiles. Adaptive control techniques using the higher performing MCUs and DSPs will allow sensors to adjust to aging effects during their operating life. Control strategies implementing computationally intense algorithms will be able to shrink a proof-of-concept laboratory demonstration into a useful solution that can have popular applications based on the increased computing capability of microcontrollers and digital signal processors.

7.1.1 Programmable Logic Controllers

Programmable logic controllers (PLCs) are used in industrial applications to control a variety of logic and sequencing processes. A typical PLC consists of a central processing unit (CPU), a programmable memory, input/output (I/O) interfaces, and a power supply [1]. Smaller, less expensive PLCs can be used and the response time and interference from noise reduced by "combining a sensor and microprocessor into one unit, commonly called a smart sensor." Functions that are performed by smart sensors in this role include

---

correcting for environmental conditions, performing diagnostics functions, and making decisions.

7.1.2 Open versus Closed-Loop Systems

The simplest control system is an open-loop system. The sensor input to a processing unit produces an output. As shown in Table 7.1, this requires basic mechanics and physics knowledge as well as mechanical-to-electrical interface knowledge for a simple system such as a fuel indicator [2]. This table shows the increasing knowledge level required to implement increasingly more complex systems with automotive applications as examples.

A closed-loop system can use (1) the sensor to modify the control strategy and (2) use a control strategy to improve the performance of the sensor [3]. Figure 7.1 shows a comparison of a capacitive surface-micromachined accelerometer using an open-loop signal conditioning circuit and a closed-loop signal-conditioning circuit. The closed-loop system has advantages over the open-loop system, including improved linearity and wider dynamic range. A comparison of several parameters is shown in Table 7.2 [3].

In general, a single-loop closed-loop system has a comparison node for a set point versus an input (the feedback signal) from a sensor regarding the status of the process that is being controlled. Based on the comparison, control action is taken to increase, reduce, or maintain speed, flow, or some other variable controlled by a mechanical actuator such as a valve or motor. The PID control algorithm is a common technique in a closed-loop system [4].

7.1.3 PID Control

Most single-loop controllers include at least three principal control actions: proportional, integral (or reset), and derivative (or rate). Each of these actions corresponds to a specific

Table 7.1 Types of Vehicle Control Systems

Name	Knowledge/Skill	Application Example
Open loop	Basic mechanics/physics, mech./elect. interface	Fuel indicator
Closed loop	Basic feedback theory	Automatic temperature control
Complex processor	Multiple function sequential processes	Engine management
Interactive system	Interactive system controls, control analysis CAE	Antilock brake & traction control
Total vehicle control	Architecture tools, failure-mode management, communication theory, structured software	Intelligent vehicle
Global	Higher level languages, simulators	External environment inputs

After: [2].

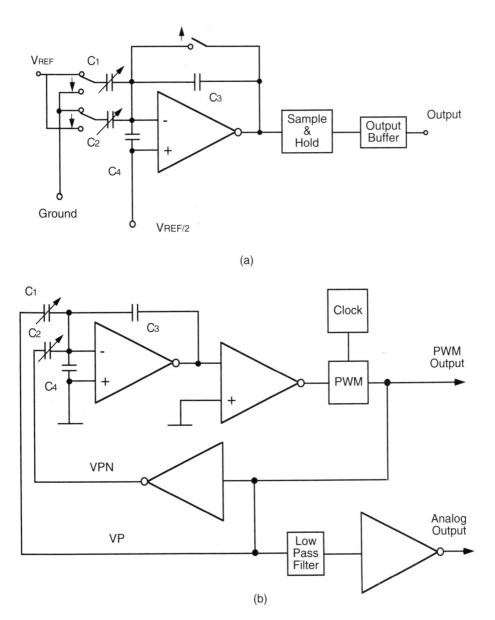

Figure 7.1 (a) Open-loop versus (b) closed-loop accelerometer circuit.

Table 7.2 Comparison of Open and Closed Loop System

Parameter	Open Loop	Closed Loop
Latch up	Possible	Not possible
Mechanical fatigue	Possible	Minimized
Linearity	Medium	High
Dynamic range	Narrow	Wide
Frequency response	Set by damping	Set by electronics
Sensitivity	Set by mechanical design	Set by electronics

technique for controlling the process set point. Proportional adjustments amplify the error by a constant amount relative to the size and sign of the deviation. The proportional band has a range of deviations, in per cent of full scale, that corresponds to the full operating range of the control element [4].

Integral control is a function of time, unlike the proportional control that has no time associated with it. Integral control responds to the continued existence of a deviation, such as an offset. The combination of proportional and integral action results in a change at a steady rate for each value of deviation. The integral rate is usually measured in repeats per minute.

Derivative control is only used in combination with the proportional or integral control action. Changes in speed and/or direction of the deviation initiate derivative control action. This action reduces the time that the system would otherwise take to reach a new set point determined by the proportional control. Controllers with derivative control have rate amplitude and rate time adjustments. Proportional-integral-derivative (PID) control can be implemented in both analog and digital systems [4].

The PID control algorithm is defined by

$$T(s) = C(s)/R(s) = 1 \tag{7.1}$$

where $T(s)$ is the desired transfer function and $R(s)$ and $C(s)$ are Laplace transforms of the reference input $r(t)$ and the controlled output $c(t)$ [5].

If $C(s)$ is not measured, the system is open loop. Otherwise, a measurement of $C(s)$ is used to provide closed-loop feedback control. Figure 7.2 shows the location of these parameters in the analog control system [5]. For $r(t)$ varying with time, a tracking problem must be solved. For $r(t)$ = constant, the problem is one of regulation. Disturbances in the forward transmission path and the feedback path can prevent $c(t)$ from ideally following $r(t)$.

The process being controlled typically dictates if only proportional, proportional and integral, or full PID should be implemented. As in most systems, the most appropriate solution is the simplest one that will achieve the desired results. Control systems based on digital techniques can improve previous analog systems. Figure 7.3 shows the simplifi-

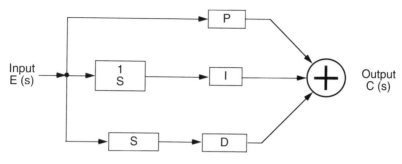

Figure 7.2 PID Control.

cation that results when a single rotary encoder provides position, speed, and commutation data in a speed-control system [6]. The number of connections to the motor is reduced from 13 to 8. The speed computation block is used to interpolate the encoder output signal for improved resolution.

7.2 STATE MACHINES

The state machine is one of the most commonly implemented functions in programmable logic [7]. The state machine is actually a sequencing algorithm. It allows an interrupt service routine to execute an algorithm that extends over many interrupts. State machines can be open or closed loop [8]. State machines are developed from a state diagram and a state table. At the beginning of the design process, the state diagram is determined from a description of the problem. A natural language describes the intended circuit function. The synthesis of the state machine for rapid and easy development of the circuit from a functional description of the state machine are areas of interest. If only a small subset of inputs is sensed while the circuit is in any given state, then an algorithmic state machine is used in place of a state diagram.

A finite-state machine is a mathematical representation of a digital sequential circuit where abstract symbols, not binary codes, represent the states. Finite-state machines can be specified in state transition tables or circuits consisting of logic gates and flip flops. These dedicated controllers can be either synchronous or asynchronous and can be implemented in hardware or software.

7.3 FUZZY LOGIC

The fuzzy logic control theory is an alternative to look-up tables and algorithm calculations that is easier to develop and has performance advantages in many applications. Fuzzy logic can be defined as a branch of logic that uses degrees of membership in sets rather than a strict true/false membership [9]. Fuzzy logic is finding acceptance in consumer, automotive,

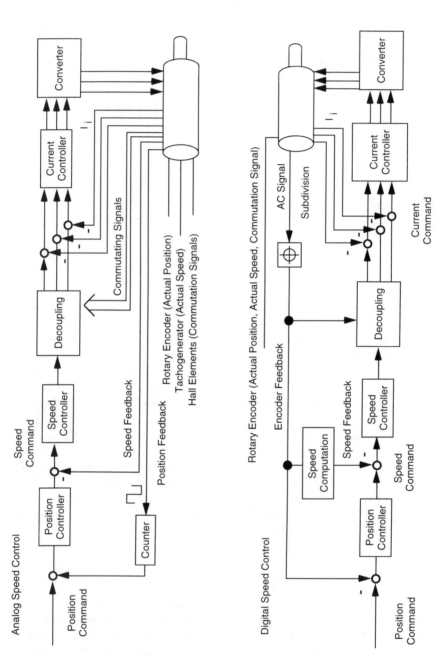

Figure 7.3 Analog versus digital speed control. (*After:* [6].)

industrial, and even decision-making applications [10]. Table 7.3 provides examples of existing applications.

In contrast to PID control, which models the system or process being controlled, fuzzy logic focuses on the human operator's behavior [11]. Fuzzy logic defines a partial, and not a strict, set membership normally associated with the Boolean logic used in MCUs and DSPs. Rules stated in natural language establish membership to the fuzzy sets.

Figure 7.4 shows an example of fuzzy logic membership for a pressure (or vacuum) input. The fuzzy set allows membership with different grades, each expressed by a number in the interval [0, 1]. In this example, membership functions and rules are established for the five pressure ranges of very low, low, normal, high, and very high. A rule for the very high pressure could be: *If* the pressure is very high, *then* decrease the power greatly, with a weight of 0.7 assigned for greatly. For low pressure, the rule could be: *If* the pressure is low, *then* increase the power slightly, with a weight of 0.3 assigned for slightly.

The final part of the fuzzy logic process is defuzzification. The defuzzification process takes a weighted average to translate the fuzzy outputs into a single crisp value.

Table 7.3 Applications of Fuzzy Logic

Consumer	Automotive	Industrial	Decision Support
VCR	Air conditioning	Extruding	Building HVAC
Camcorder	Transmission control	Mixing	Medical diagnostics
Television	Suspension control	Furnace control	Transportation
Vacuum cleaner	Cruise control	Temperature control	Elevator control
Clothes dryer		Conveyor control	
Washing machine			
Hot water heater			

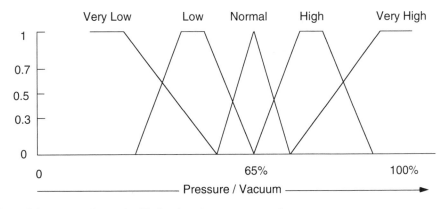

Figure 7.4 Fuzzy logic membership functions for pressure control.

Output membership functions are typically defined by singletons. A singleton in an 8-bit system is defined by an 8-bit value that represents the output value corresponding to the linguistic label of the output system.

To simplify the implementation of fuzzy logic on existing microcontrollers, a fuzzy kernel can be developed by the MCU manufacturer [12]. A fuzzy kernel, or engine, is software that performs the three basic steps of fuzzy logic: fuzzification, rule evaluation, and defuzzification. As shown in Figure 7.5, the fuzzy inference unit (FIU) receives system inputs and information from the knowledge base for each step of the process. All application-specific information is contained in the knowledge base, which is developed independently from the fuzzy inference program.

With fuzzy logic, applications that were once thought to be too complex to be practical are being easily controlled. Furthermore, the rules in a fuzzy system often hold true even if the operating parameters of the system change. This is typically not true in conventional control systems.

In general, fuzzy logic does not require special hardware. The software requirements can be low, based on a limited number of rules to describe the system. For most control tasks, software running on standard processors can perform fuzzy logic operations. Specialized hardware does simplify code development and boost computational performance for low-cost applications. Fuzzy logic allows 8-bit MCUs to perform tasks that may have otherwise required a more powerful MCU such as a RISC processor.

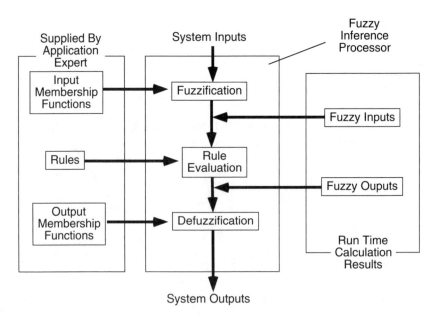

Figure 7.5 Fuzzy software inference unit.

Fuzzy logic has been used with multiple sensors to produce a mobile security robot for evaluating the environment in an art gallery [13]. As shown in Figure 7.6, the fuzzy logic certainty engine in the robot weighs multiple variables, creates a composite picture from data supplied by the sensors, and performs real-time threat identification and assessment. The robot is able to analyze data based on the location (of the art), the time of day, historical trends, and external conditions. A wide range of safety threats detrimental to art objects have been identified by the robot during a 36-month evaluation. Air quality assessment for conditions harmful to humans is an obvious extension of this technology.

7.4 NEURAL NETWORKS

Unlike fuzzy logic, neural networks allow the system, instead of an expert, to define the rules. By definition, a neural network is a collection of independent processing nodes that solve problems by communicating with one another in a manner roughly analogous to neurons in the human brain [11]. Neural networks are useful in systems that are difficult to define. They have an additional advantage of the ability to operate in a high-noise environment. Complex or numerous input patterns are among the problems that neural networks are addressing [14].

The structure of a neuron is shown in Figure 7.7(a) [14]. The neuron learns from a set of training data. Each input to the neuron is multiplied by the synapse weight. The neuron sums the results of all the weighted inputs and processes the results with a typically

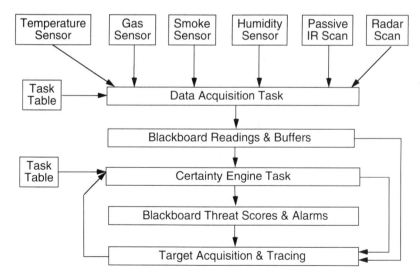

Figure 7.6 A variety of sensors provide input to the fuzzy logic certainty engine of a mobile robot. (*After:* [13].)

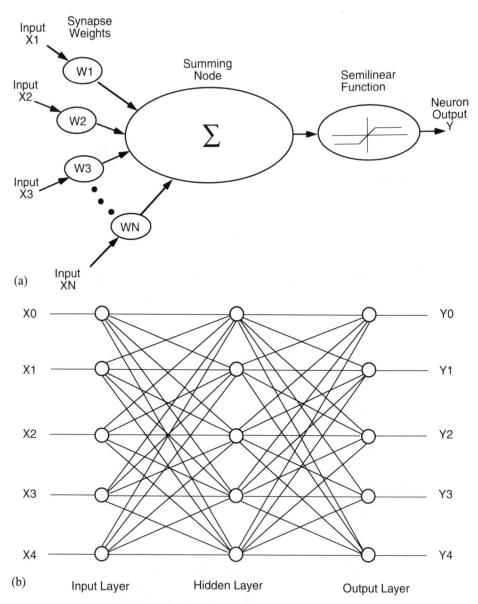

Figure 7.7 (a) Neuron and (b) neural network. (*After*: [14].)

nonlinear transfer function to determine the output. A local memory stores previous computations and modifies the weights as it learns.

A neural network consists of a number of neurons connected in some manner, as indicated in Figure 7.7(b). In this model, a hidden layer exists that prevents direct interaction between the input and the output. However, the hidden layer is not required and more than one may exist. Also, the number of neurons does not have to be the same in each layer. To train or condition the neuron for proper response, several sets of inputs with known results are applied to the neural network. The output is compared for each set of inputs to the known results, and the weights are adjusted to compensate for errors.

An example of a potential application for a neural network is the control of an automotive fuel injection system [15]. Production vehicles meet current emissions regulations using calibration and look-up tables. However, emission regulations for 2003 have further reductions of HC, NOx, and CO emissions that may require the function approximation, learning, and adaptive capabilities of neural networks. By using a neural network control, a stoichiometric air-fuel ratio (A/F) can be maintained over the life of the vehicle even if the engine dynamics change. Experimental results with a neural network and a linear A/F sensor have demonstrated the capability to control stoichiometry within a ±1%, which was better than the production control unit.

7.4.1 Combined Fuzzy + Neural

Fuzzy logic and neural networks are being combined to utilize the best features of each technology. One approach begins with a set of fuzzy rules that have been well-tuned by an expert using trial-and-error methods. A neural-like adaptive mechanism is then installed in the fuzzy system to handle exceptional circumstances after the system is in use. These systems compensate for load variables and for wear that occurs over time [16].

In an alternate approach, the fuzzy system is coarsely defined by experts. The fuzzy rule base is then refined with a neural network. The neural network adapts to minimize errors.

Another approach combines fuzzy associative memories with neural networks. In general, an associative memory is a neural architecture used in pattern recognition applications. The network associates data patterns with specific classes or categories it has learned. This combination produces a system in which the neural network front-end learns rules from training data and then supplies those rules to a fuzzy logic back-end to execute the rules [16].

The previous examples started with a fuzzy system and applied neural network learning. Other researchers have started with neural networks and applied fuzzy logic. In these systems, the network adapts in a more intelligent manner. The ultimate technology in this area could be evolutive learning—the application of genetic algorithms to combine fuzzy and neural systems automatically. Genetic algorithms are guided stochastic search techniques that utilize the basic principles of natural selection to optimize a given function

[17]. These advanced control concepts are being explored for process optimization in manufacturing complex ICs.

7.5 ADAPTIVE CONTROL

Adaptive control is a system of advanced process-control techniques that is capable of automatically adjusting or adapting to meet a desired output despite shifting control objectives and process conditions or unmodeled uncertainties in process dynamics [17]. Because of their capacity to learn, neural networks have the capability to provide adaptive control, but they are not the only means. As shown in Figure 7.8, the difference between the desired performance and the measured performance in the control system is the starting point for the adaptive process [18]. The adaptive system modifies the parameters of the adaptive controller in order to maintain the performance rating close to the desired value.

All adaptive control systems have a conventional servo-type feedback loop and an additional loop designed to identify the online process and determine the parameters to be adjusted on the basis of the process parameters. The addition of an identification algorithm and an outer control loop maintains performance in the presence of disturbances. An indirect adaptive scheme has an additional outer loop to perform system identification and it provides access to process parameters for monitoring and diagnostics [18].

Adaptive control can be obtained in fuzzy logic systems using a controller that operates at a higher level to perform tuning and adaptive control functions [19]. The architecture for this approach is compared to a low-level direct closed-loop system in Figure 7.9. Adjustments that the fuzzy logic controller performs include online and offline tuning of the controller parameters, online adaptation of the low-level controller, and self-organization and dynamic restructuring of the control system.

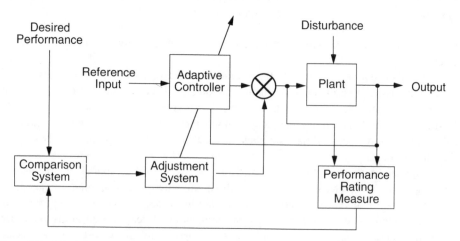

Figure 7.8 Adaptive control.

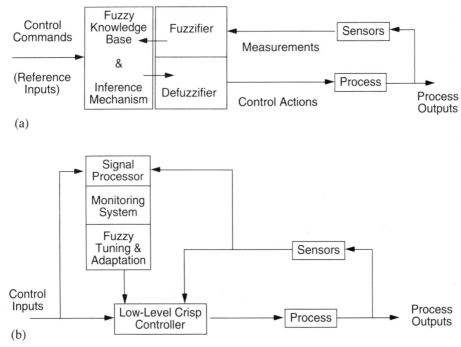

Figure 7.9   Low-level direct control (a) and high-level tuning/adaptive control (b) in fuzzy logic control system. (*After*: [19].)

7.6 OTHER CONTROL AREAS

Computationally intense models such as Hidden Markov Modeling (HMM) used in automatic speech recognition will be used more frequently as available computing power increases [20]. Higher performance for real-time control is achieved in digital signal processors (DSP) and multiprocessor systems. For example, automatic speech recognition (ASR) has been implemented in ROM on a DSP chip. The HMM technique treats speech as a stochastic process for matching the input to a word command. In these systems, sensor input from a microphone directly affects the signal to noise ratio of the input. Limited bandwidth and high-noise systems such as telephone lines still cause problems to some existing speech recognition systems. However, continuing improvements in algorithms and the computational engine are improving the process.

Multiprocessor computing is used today for controlling industrial robotics. These powerful processors support the implementation of machine vision, speech recognition, and optical character recognition. Techniques used in multiprocessing include common Fourier transform, convolution, linear prediction, dynamic time warping, template matching, and spectral processing [21].

The discrete Fourier transform is used to determine the frequency components of a time series. Multiplying a matrix of cosine and sine functions with the time-series vector yields the frequency-series vector. By computing a 64-element time-series vector concurrently on 64 processors (one component per processor), processing time was reduced to 1/64 of the uniprocessor's time.

Convolution is used for filtering a time series. It requires calculating the inner products of time-series sections with a coefficient vector that defines the filter characteristics. By using a 16-element convolution kernel or software function and distributing the elements in a 128 processor with only one broadcast of the time-series vector, the computation was performed 128 times faster.

Linear prediction is used to obtain smoothed, compact spectral representations of a signal. It requires calculating the autocorrelation function for the time signal. This is identical to convolution calculation with the convolution kernel replaced by a section of the time series for autocorrelation. By processing one coefficient per processor, the autocorrelation function is calculated N times faster, where N is the number of autocorrelation coefficients computed. Linear predictive coding is frequently used for speech recognition.

Dynamic time warping calculates the similarity between two signals that have differing nonlinear time-alignment patterns. The computation determines the optimum path through a matrix of distances formed from the squared Euclidean distances between discrete sample sections of the signals. Multiple paths are explored based on the selected algorithm, and the minimum total distance defines the similarity and best alignment path. Template matching in linear-vector quantization classifier requires a similar calculation of the Euclidean distance between and unknown vector of features and a lexicon of templates. Time warping is another approach for speech recognition.

Spectral processing involves a sequence of computations. The computations include a weight vector, Fourier transform, conversion from real to imaginary components to power, and conversion from power to decibels. The final calculation of the power components of the frequency spectrum is converted to decibels using a standard linear-to-logarithmic library function. This amount of processing is indicative of speech and image recognition.

Associative memory is used in pattern recognition applications, in which the network is used to associate data patterns with specific classes or categories it has already learned [17].

The control areas discussed in this section are aimed at some of the high-end computing requirements in sensory systems. As shown in Table 7.4, these technologies are essential for many smart-sensing applications. More than one approach can be used to implement a solution for these complex systems.

7.6.1 RISC versus CISC Architecture

The complex instruction set computer (CISC) architecture is the established approach for desktop computing and embedded control applications. An embedded system has one or

Table 7.4 Emerging Techniques for
Smart-Sensing Applications

Technique	Enabling Technology
Voice recognition	DSP, NN, RISC, MP
Handwriting recognition	DSP, MP, NN, FL
Multimedia	DSP, FL, RISC, MP
Digital sound	DSP
FAX/modem	RFICs, CISC
Data storage	Flash, DRAM, SRAM
CD-ROM	CISC
Virtual reality	DSP, RISC, FL, MP
Synthesized speech	DSP, RISC, NN, MP
Data compression	DSP
Pattern recognition	DSP, hi-perf MCU, NN

MP = Multiprocessing
FL = Fuzzy logic
NN = Neural network

more computational devices (which may be microprocessors or microcontrollers) that are not directly accessible to the user of the system. A relatively new addition for computing requirements is the reduced instruction set computer (RISC) architecture. RISC architecture is being used by computer and workstation manufacturers who are seeking the highest performance levels. Embedded computing applications are also utilizing the performance that this approach provides. The transition from CISC to RISC can be understood by analyzing the following equation for microprocessor performance:

$$\text{Performance} = \frac{\text{MHz} \cdot \text{Instructions/Clocks}}{\text{Instructions}} \qquad (7.2)$$

CISC-based architectures allocate transistors to decreasing the denominator of (7.2) [22]. However, based on recent improvements in price-performance of semiconductor memory and significant advances in compiler technology, transistors can be reallocated to increasing the numerator. This is the basis of the RISC architecture, which frees up transistors for integrating additional components. The additional components can be peripherals or microcoded kernels that improve the performance in a specific application.

Two approaches, CISC and RISC, are now available to solve a variety of application problems in several markets. Since no single computing solution is correct for all applications, designers must choose the approach that satisfies their design criteria. For a given application, there is a ''best fit'' solution and more than one feasible solution. CISC-based designs will continue to satisfy a significant number of applications, especially in embedded control. However, RISC has established a credible presence in a range of commercial computing environments and its use will expand in embedded applications. RISC architecture in an embedded environment does not take advantage of the 64-bit bus, zero-wait-state memory, or the amount of memory that is integrated on one desktop MPU chip. To be affordable, the RISC performance is compromised but still beyond the capabilities of today's CISC MCU.

CISC microcontrollers range from simple 4-bit units and the most popular 8-bit units up to 32-bit units. The lower cost versions are easily combined with single sensors, such as a pressure, flow, or acceleration sensor, to provide a smart sensor. The high-performance 32-bit CISC and RISC units are allowing complex strategies and diagnostics in automotive engine and powertrain control systems. They also are required for more complex sensing in imaging and navigation systems.

7.6.2 Combined CISC, RISC, and DSP

An interesting demonstration of the value of both CISC and RISC is provided in a communications control circuit shown in Figure 7.10 [23]. This is a single-chip solution incorporating CISC, RISC, *and* DSP technology. CISC technology (68000-based) is used for the integrated multiprotocol processor. Three serial communications channels handle HDLC, asynchronous (UART), BYSYNC, and synchronous protocols. A PCMCIA block

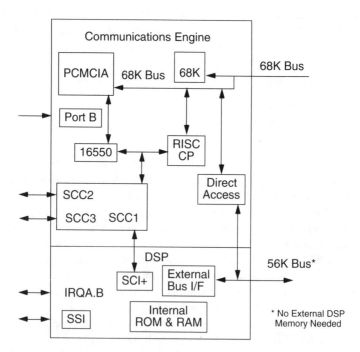

Figure 7.10 Block diagram of combined CISC, RISC, and DSP for data communications.

provides a direct interface to a PCMCIA bus (PC card). RISC is used for the communication processor. As a result, the ports operate independent of the 68000 host processor and allow the 68000 to handle higher level control tasks [24]. DSP technology provides increased performance through the parallel (Harvard) architecture. This allows it to simultaneous fetch program, X-data, and Y-data memory. Since both cores use common memory, cost is reduced. The communication processor can be applied to sensing applications that include voice compression in tape answering machines and simultaneous voice/data transmission. It demonstrates the extent that technologies can be integrated and indicates the potential for sensor control integration in the future.

7.7 IMPACT OF ARTIFICIAL INTELLIGENCE

Fuzzy logic and neural networks will undoubtedly affect not only the control systems, but also the sensors that are used in these systems. One analysis, shown in Figure 7.11, suggests that artificial intelligence will require sensors with "lower" accuracy and, subsequently, will cost less [25]. At the same time, the number of sensor applications will increase as a result of artificial intelligence (and the lower cost). Artificial intelligence includes the newer control techniques, such as fuzzy logic and neural networks.

The control decisions made in fuzzy logic systems can be made in spite of absolute accuracy from the sensory data [26]. As a result, performance specifications will change

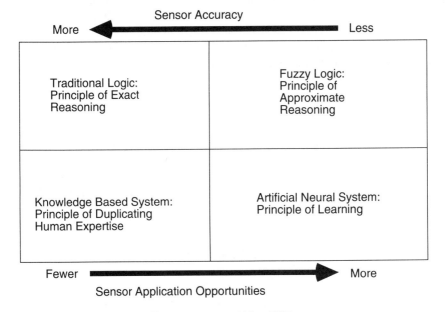

Figure 7.11 Impact of artificial intelligence on sensors. (*After*: [25].)

from point-to-point to domain-type specifications. Domain-type specifications will allow the sensor to function with wider tolerance in terms of repeatability, accuracy drift, and linearity. A natural extension of the fuzzy logic rules to the sensor has the sensor expressing the measurement in terms of grade of membership to the predefined domain. These principles have already been applied to color, proximity, fluid control, and position sensors.

An extensive analysis of classic control theory, fuzzy logic, neural networks, and rough set theory has been performed for smart sensors [27]. The authors have defined an operator's inference model that allows the evaluation of the situation, assignment of the situation to a characteristic state, and the selection and execution of the proper characteristic control. The model has been used to evaluate smart sensors and control system design.

As part of their analysis, the rough set approach has been analyzed. The rough set is similar to fuzzy logic, except that a rough ADC replaces the fuzzifier in the controller, and a rough DAC replaces the defuzzifier. The inference engine in the controller evaluates the decision table for the rough set. The decision tables used to derive the rules in the rough controller are simple to understand and easy to edit.

As part of their findings, the authors determined that classic control theory, although a mature technology with widely supported hardware, accomplishes increasing more complex computing by increasing the MIPS (millions of instructions per second). Fuzzy logic has commercial developments available that are making its acceptance widespread, but it does have safety and reliability disadvantages. The rough set can be used for pattern matching and is fast and low-cost, but does have stability and completeness concerns similar to fuzzy logic and neural networks. Neural networks can possibly provide an all-analog system, which is easily integrated with sensors, but they have an added problem of an unknown decision basis. However, attention to the areas of concern is allowing all of these systems to be evaluated for new applications. The combination of these control techniques with sensors will achieve new levels of smart sensing.

7.8 SUMMARY

Control techniques and significantly improved computing capabilities have been discussed in this chapter. Fuzzy logic is already finding broad acceptance in many control applications. Complex systems will benefit from fuzzy logic and other advanced approaches. However, the sensors in these systems may not have to be as accurate, and therefore could offset at least some of the added cost of getting smarter.

REFERENCES

[1] "Controls and Sensors," *Power Transmission Design*, Jan. 1995, A179–A190.
[2] Gormley, J., and D. A. Mac Isaac, "The Systems Approach (Better Late Than Not at All)," *Proc. of Convergence '88*, Dearborn, MI, Oct. 17–18, 1988, pp. 107–118.

[3] Dunn, W., and R. Frank, "Automotive Silicon Sensor Integration," SAE SP-903 *Sensors and Actuators 1992*, SAE International Congress & Exposition, Detroit, MI, Feb. 24–28, 1992, pp. 1–6.

[4] Wachstetter, J., "Understanding the Principal Control Actions of Single-Loop and Multiloop Controllers," *I&CS*, Aug. 1994, pp. 45–51.

[5] Stokes, J., and G.R.L. Sohie, "Implementation of PID Controllers on the Motorola DSP56000/DSP56001," *Motorola APR5/D, Rev. 1*, 1993.

[6] Hagl, R., and S. Bielski, "Rotary Encoders Make Digital Drives Dynamic," *Machine Design*, Aug. 22, 1994, pp. 85–93.

[7] Hil, F. R., *Computer Aided Logical Design with Emphasis on VLSI*, New York, NY: John Wiley & Sons, Inc., 1993.

[8] Preston, J. V., and J. D. Lofgren, "FPGA Macros Simplify State Machine Design," *Electronic Design*, Dec. 5, 1994, pp. 109–118.

[9] Self, K., "Designing with Fuzzy Logic," *IEEE Spectrum*, Nov., 1990, pp. 42–44, 105.

[10] Lewis, M., "Fuzzy Logic: A New Tool for the Control Engineer's Toolbox," *Power Transmission Design*, Sept. 1994, pp. 29–32.

[11] Schwartz, D. G., and G. J. Klir, "Fuzzy Logic Flowers in Japan," *IEEE Spectrum*, July 1992, pp. 32–35.

[12] Sibigtroth, J. M., "Fuzzy Logic for Small Microcontrollers," *Wescon/93 Conference Record*, San Francisco, CA, Sept. 28–30, 1993, pp. 532–535.

[13] Holland, J. M., "Using Fuzzy Logic to Evaluate Environmental Threats," *Sensors*, Sept. 1994, pp. 57–60.

[14] Wright, M., "Neural Networks Tackle Real-World Problems," *EDN*, Nov. 8, 1990, pp. 79–90.

[15] Majors, M., J. Stori, and D-I. Cho, "Neural Network Control of Automotive Fuel-Injection Systems," *IEEE Control Systems*, June 1994, pp. 31–35.

[16] Johnson, R. C., "Gap Closing Between Fuzzy, Neural Nets," *Electronic Engineering Times*, April 13, 1992, pp. 41, 44.

[17] May, G. S., "Manufacturing ICs the Neural Way," *IEEE Spectrum*, Sept. 1994, pp. 47–51.

[18] Renard, P., "Implementation of Adaptive Control on the Motorola DSP56000/DSP56001," *Motorola APR15/D*, 1992.

[19] De Silva, C., and T-H. Lee, "Fuzzy Logic in Process Control," *Measurement and Control*, June 1994, pp. 114–124.

[20] Quinell, R. A., "Speech Recognition: No Longer a Dream but Still a Challenge," *EDN*, Jan. 19, 1995, pp. 41–46.

[21] Skinner, T., "Program a Multiprocessor Computer," *Electronic Design*, Jan. 23, 1995, pp. 57–66.

[22] Krohn, N., R. Frank, and C. Smith, "Automotive MPU Architectures: Advances & Discontinuities," *Proc. of the 1994 International Congress on Transportation Electronics*, SAE P-283, Oct. 1994, Dearborn, MI, pp. 71–78.

[23] "MC68356 - Signal Processing Communication Engine," Motorola Product Brief.

[24] Bursky, D., "No-Compromise Controller Combines DSP, Data Comm," *Electronic Design*, June 13, 1994, pp. 79–88.

[25] Abdelrahman, M., "Artificial Intelligence Expands Sensor Applications," *EC&M*, Dec. 1991, pp. 20–23.

[26] Abdelrahman, M., "Fuzzy Sensors for Fuzzy Logic," *Control Engineering*, Dec. 1990, pp. 50–51.

[27] Maiten, J., A. Mrozek, R. Winiarczyk, and L. Plonka, "Overview of Emerging Control and Communication Algorithms Suitable For Embedding into Smart Sensors," *Proc. of Sensors Expo*, Cleveland, OH, Sept. 20–22, 1994, pp. 485–500.

Chapter 8

Transceivers, Transponders, and Telemetry

"Tricorder readings indicate an unusually high level of nitrium."
—Lt. Commander Data, from *Star Trek: The Next Generation*

8.1 INTRODUCTION

One of the more monumental changes occurring at the end of the 20th century is also affecting sensing; that is, the desire for portable, wireless products. The proliferation of portable computers combined with wireless communication will have the same effect on sensing that the shift from mainframe to personal computers had on the factory floor. The number of sensed parameters and ease of obtaining measurements will increase based on remote, wireless sensing. Remote measurements require a combination of the digital technologies, which were discussed in Chapter 5, and radio frequency (RF) semiconductors including radio frequency ICs (RFICs). In some instances, RF technology is used to perform the sensing function.

The use of RF technology in remote sensing has historically been associated with geophysical analysis of parameters such as air and surface temperature, wind velocity, and precipitation rate. This information is necessary to predict weather and is typically gathered from equipment mounted on aircraft or satellites. These applications of RF continue to grow; however, newer commercial applications that are portable and have much lower cost objectives will be the focus of this chapter.

Portable wireless products require low power consumption to increase their useful life before requiring battery replacement. For low power consumption measurements, high-impedance sensors, sleep mode, and smart techniques such as the pulse-width modulation (PWM) technique that was discussed in Chapter 5 are required. Depending on the type of measurement (static or dynamic) and how frequently it must be made, periodic readings can be transmitted to distant recording instruments in process controls, hazardous material monitoring systems, and a variety of mobile data acquisition applications. These readings would have been more time consuming, dangerous, prohibitively expensive, or difficult with previously available technology.

8.1.1 The RF Spectrum

The electromagnetic spectrum that is available for wireless communications in the United States has been expanded based on recent legislation. The Federal Communications Commission (FCC) regulation Title 47, part 15 covers unlicensed security systems, keyless entry, remote control, and RF local area networks (LANs). Table 8.1 identifies key

Table 8.1 RF in the Electromagnetic Spectrum

RF Spectrum	Frequency	Miscellaneous
Industrial/scientific/medical	902–928/ 2,400–2,483.5/ 5,725–5,850 MHz	
Microwave radar	10.25, 24, or 34 GHz	
Radar proximity sensor	1 MHz	
SAW in tolls	856 MHz	
Remote key	41 or 230 MHz	
WLAN	2.4 GHz	*DTR ≥ 1.6 Mbps
Altair™ WLAN	18 GHz	DTR = 15 Mbps
LAN PCMCIA card	900 MHz	
RF-ID	100 kHz to 1.5 MHz and 900 MHz to 2.4 GHz (915 MHz)	
GPS (L1 & L2)	1,575.42, 1,227.60 MHz	
Ultrasonic position sensing	20–200 kHz	—
Piston temperature telemetry	1 MHz	
Tire monitoring system	355 MHz or 433.92 MHz	
Medical telemetry (UHF)	450–470 MHz	
Medical telemetry (VHF)	174–216 MHz	

Band	Microwave Electronics Band Frequency	Wavelength
L	1–2 GHz	30.0–15.0 cm
S	2–4 GHz	15.0–7.5 cm
C	4–8 GHz	7.5–3.7 cm
X	8–12.5 GHz	3.7–2.4 cm
Ku	12.5–18 GHz	2.4–1.7 cm
K	18–26.5 GHz	1.7–1.1 cm
Ka	26.5–40 GHz	1.1–0.75 cm
mm	40–100 GHz	0.75–0.3 cm
Ultraviolet		0.1–0.4 μm
Visible		0.4–0.7 μm
Infrared		0.7–0.1 cm

*DTR = data transfer rate

frequency ranges that are already the focus of several industries. The industrial, scientific, and medical (ISM) bands are being used for several applications. Integration is the key to higher performance, smaller packages, and lower cost in the RF arena, as well as previously discussed systems, with several technologies used in various frequency ranges.

Several technology choices exist in the RF front-end of a portable communication product, as shown in Figure 8.1 [1]. Silicon competes with GaAs (gallium arsenide) in the 1 to 2 GHz (gigahertz, 10^9 Hz) range; however, in the 2 to 18 GHz range, GaAs is the only solution. With a supply voltage of 3V, high-frequency (1 GHz) operation of GaAs has an efficiency of 50% versus silicon bipolar's 40% or lateral diffused MOS's (LDMOS's) 43%. The tradeoff in efficiency versus cost must be evaluated prior to initiating a custom design. The downconverter and low-noise amplifier (LNA)/mixer, transmit mixer, antenna switch, driver and ramp, and power amplifier can be designed using RF chip sets consisting of different IC technologies. The chip set approach can be a cost-effective alternative to higher levels of integration. High-frequency designs for radio frequency are considerably different than high-speed digital processes. The frequency range in RF circuits of 800 to 2,400 MHz (megahertz, 10^6 Hz) is much higher than the fastest digital circuits. Mixed-signal technology is required that combines not only digital and analog, but also RF. Circuit isolation is required to prevent unwanted coupling of signals, which can range from 3 V_{p-p} to less than 1 μV_{p-p}. RFICs are being designed for a number of high-frequency applications using a variety of technologies, depending on the application. The term MMIC, for monolithic microwave integrated circuit, is used for RFICs in the microwave range.

The transition from RF to digital baseband is simplified by mixed-signal, analog-digital IC's. As shown in Figure 8.1, a phase-locked loop (PLL) synthesizer and prescalar as well as a modulator/demodulator are required for this transition. PLL frequency synthesizers are required in RF applications to provide digital tuning capability for wireless digital communication to imlement a cost-effective multiple-channel design. A PLL allows a precise stable frequency to be generated with an MCU controlling the frequency. Advanced CMOS processes provide the lowest power dissipation and a low operating voltage for 100 to 300 MHz. BiCMOS increases the frequency range (to about 2 GHz) and provides flexible circuit elements at moderate power dissipation. An advanced bipolar process, like the MOSAIC™ (Motorola Self-Aligned IC) process provides low power dissipation at a lower cost and has an even higher frequency range than BiCMOS. GaAs provides the lowest operating voltage at the highest cost, but can operate above 2 GHz.

Mixed analog and digital process technology is required for the adaptive differential pulse code modulation (ADPCM) codec (coder/decoder) shown in Figure 8.1 that translates analog into a digital format in personal communication devices. The digitized signal is transmitted over the RF channel. The ADPCM is designed for 32 Kbps, which is the codec technology for many personal communication systems worldwide, as well as 64-, 24-, and 16-Kbps data rates. Digital signal processor (DSP) technology is used to achieve optimum data rate and RF bandwidth. Precision analog and high-performance digital technologies are required for the transition from analog to digital regime and back.

GaAs semiconductors are being designed to address the highest speed applications

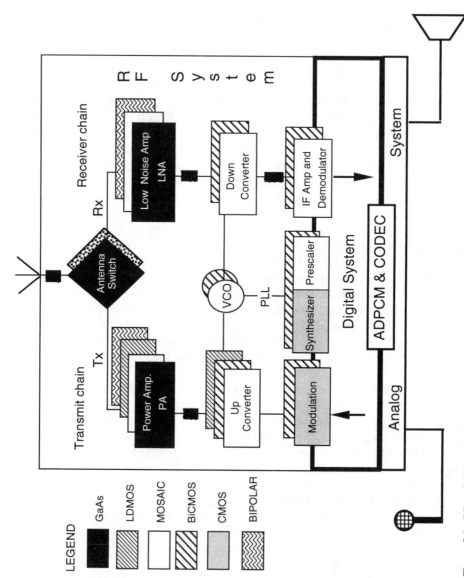

Figure 8.1 RF to digital transition.

and are well-suited to low-voltage operation. However, they are more expensive than silicon and should only be used where their performance would be difficult or impossible to achieve with silicon. The use of GaAs technology adds another dimension to circuit partitioning that must be considered for RF circuits.

8.1.2 Spread Spectrum

Spread spectrum is one of the secure methods of transmitting information via radio signals that has recently (1991) expanded from military use to commercial use. The spread-spectrum signal is of a much lower magnitude but is generated by a much broader frequency range, 26 MHz compared to a narrowband signal of less than 25 kHz, as shown in Figure 8.2. Narrowband transceivers (combined transmitter and receiver) are highly susceptible to interference by frequency source near their carrier frequencies. Since they operate within the audio range, the equipment being monitored frequently generates the interference [2].

Spread spectrum offers improved interference immunity, low interference generation, high data rates, and nonlicensed operation at practical power limits [3]. To conform to FCC guidelines, the total transmitted power must be 1W maximum and the spectral density (the power at any specific frequency) must be no greater than 8 dB (decibels) in a 3-kHz

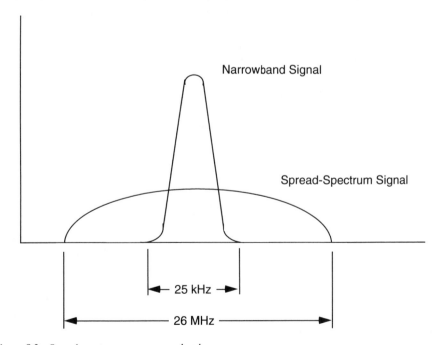

Figure 8.2   Spread spectrum versus narrowband.

bandwidth. Two techniques are used to distribute the conventional narrowband signal into a spread-spectrum equivalent: direct sequencing and frequency hopping.

Direct sequence is a spread-spectrum approach that creates an instantaneous RF signal consisting of many frequencies spread over a portion of the frequency spectrum. The information signal, which is usually digital, is combined with a much faster stream of pseudorandom binary code that is repeated continuously. This technique is difficult to detect, and recapturing the information requires duplicating the spreading code at the receiver end.

Another spread-spectrum approach is frequency hopping. Both base and subscriber, or handset and base, "hop" from frequency to frequency in a simultaneous fashion [4]. The theory states that noise tends to occur at different frequencies at different times. Therefore, even though a part of the transmission may be lost due to interference, enough of the message will be received by hopping through the interference to create a noticeably better output when compared to fixed-frequency systems. A variety of applications and their frequencies are shown in Table 8.1. The 1.8- to 2.4-GHz band is also open for data collection in Europe, which at present is the second largest market for RF data.

Spread spectrum's ability to put a greater number of carriers in a specific bandwidth and increase the security of the transmitted data is enabling several new wireless applications for sensing. Spread-spectrum techniques combined with new frequency ranges in the unlicensed ISM bands have made new applications possible *and* easy to implement. In addition to communication, commercial applications of spread spectrum include wireless LANs, security systems, instrument monitoring, factory automation, remote bar-code reading, automatic vehicle location, pollution monitoring, medical applications, and remote sensing of seismic and atmospheric parameters.

8.2 WIRELESS DATA AND COMMUNICATIONS

Error-free wireless data is becoming possible due to digital data transmission. A number of protocols are vying for acceptance in RF signal transmission. Existing access methods include time-division multiple access (TDMA), frequency-division multiple access (FDMA), and (the newest) code-division multiple access (CDMA). Users are assigned a time slot, frequency, or code that allows multiple users to share available frequency spectrum. CDMA provides an additional dimension and option that allows more efficient utilization of the available spectrum and therefore greater capacity (up to 20 times greater) with a given bandwidth [5]. This means an increased number of users can access the system and maintain the level of service (access, data rate, bit error rate (BER), etc.) with only a slight degradation in process gain. The discontinuous nature of human speech is the key to one aspect of CDMA technology. Only one-third of the air time is used for speech, and the rest is spent on pauses between words, syllables, thoughts, and listening. CDMA uses a variable-rate vocoder (voice encoder) that performs a speech-encoding algorithm to transmit speech at the minimum data rate needed. Encoding rate is from 1 to 8 Kbps, depending on speech activity.

Several techniques that include frequency shift keying (FSK), amplitude shift keying (ASK), on-off keying (OOK), differential phase shift keying (DPSK), quadrature phase shift keying (QPSK), and Gaussian minimum shift keying (GMSK) are used to modulate the signal transmission. Techniques previously used for military and satellite communications, such as differential quadrature shift keying (DQSK) and GMSK, maximize the information density in a spectral bandwidth. The choice depends on cost and performance objectives within the system.

Cellular digital packet data (CDPD) technology divides data into small packets that can be sent in short bursts during the idle time between cellular voice transmissions. This open specification is designed to support a variety of interoperable equipment and services and is backed by major cellular carriers. CDPD allows faster transmission (19.2 Kbps) of short messages compared to the existing circuit-switched system, which handles longer files [6]. CDPD also incorporates error-correction protocols. Some of the competing protocols are listed in Table 8.2. Several RF-specific terms are defined in the glossary.

8.2.1 Wireless Local Area Networks

Wireless local area networks (WLAN) operating at 18 GHz are able to propagate well within a building, yet diffuse rapidly outside. FCC licensing is required for WLANs in the digital termination service (DTS) band, 18 to 19 GHz, which includes the use of low-power radio communications within buildings. Each of the 10-MHz channels within this band can deliver 15 Mbps [7]. Some WLANs also operate in the lower frequency range of 900 MHz to 6 GHz. In addition, there are also a variety of infrared links for wireless networks. These indoor systems have a range of 100m and data rates of 1 Mbps.

8.2.2 Fax/Modems

Real-time sensing using a Personal Computer Memory Card International Association (PCMCIA) card fax/modem installed in any portable computer establishes a flexible, easily installed link to a base computer. Up to 28,800 bps are already possible using available technology. PCMCIA cards are also being designed specifically to handle sensor data [8].

Table 8.2 RF Communication: Wireless LAN Protocols and Sponsoring Agency

Protocol	Sponsor(s)	Comments
X.25 packet protocol	CCITT	Wireless extension of land-line specification
802.11	IEEE	Proposed standard, uses ISM frequencies
CDPD	McCaw	Uses unused capacity on cellular voice
RAM Mobitex	Ericcson	Message packet data system
ARDIS™	Motorola & IBM	Data packets at either 9.6 or 14.4 Kbps

One device provides the modem function and a level of intelligence that can obtain information from sensors located in remote or mobile devices.

An error-correcting protocol, derived from the asynchronous balance mode of the Consultive Committee of the International Telephone and Telegraph (CCITT)-defined X.25 packet protocol retransmits only those frames that are not recognized by the receiver (refer to Figure 8.3). The modulation technique easily synchronizes transmitted data. A basic high-level data-link control (HDLC) concept and non-return-to-zero, inverted (NRZI) bit patterning increases the clock-recovery data for the receiver. Data is transmitted in packets of up to seven adjustable-length frames. A compression algorithm built into the firmware kernel allows data compression, transfer, and analysis prior to transmission. Error-correction capability is achieved by allowing the receiver to store ahead frames that are received out of sequence and request the retransmission of any missing frames.

A file-level protocol facilitates the exchange of data between the mobile unit and the base unit. A microprocessor controls the data transmission and services the serial ports and parallel port. Global position satellite (GPS) data (see Section 8.3.3), mobile data terminal, and bar-code readers/scanners are among the immediate applications for this RF data transmission.

8.2.3 Wireless Zone Sensing

Wireless zone sensors (WZS) have been developed for buildings using the spread-spectrum technique. The wireless sensors are part of the modern building automation systems designed for energy efficiency and precise control of many individual areas [9]. Figure 8.4 demonstrates a typical system with the wireless zone sensors.

The WZS is a spread-spectrum transmitter that provides room temperature and other status information to a receiver located up to 1,000 feet away. A translator in this unit converts the wireless data to a wired communication link. The wired communication link distributes data to variable air volume (VAV) terminal controllers and to the building management system. The VAV adjusts the valve opening to allow more or less flow to the area as required by the WZS. A sensor is matched to its associated VAV by a setup tool

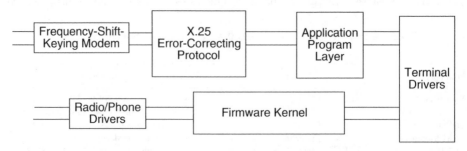

Figure 8.3 Wireless data using PC card.

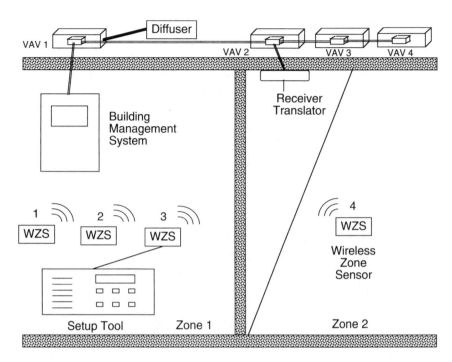

Figure 8.4 Wireless zone sensing.

that is connected to each WZS only during the installation process. The lithium-battery-powered WZSs are designed to operate an average of two years without changing batteries.

Wireless data loggers are being used in clean rooms to pinpoint localized humidity fluctuations [10]. In this wireless sensor application, the measurement is only required to locate the source of a problem. Flexibility in choosing the test sites as well as avoiding clean-room suits and drilling holes through the wall were paramount design considerations. Again, the low interference possible using spread-spectrum technology has provided the wireless transmission of real-time data in an environment that previously prohibited radio transmission. Up to 20 data loggers have been linked to a single wireless base station. As a result, a single PC can handle the measurements from a number of widely separated data collecting units.

8.2.4 Optical Signal Transmission

The infrared spectrum is also used for transmitting data. It is low-cost for cost-sensitive applications and does not require licensing. However, it is limited to line of sight (LOS) transmission and cannot penetrate solid objects. The infrared spectrum (indicated in Table 8.1) is from 0.7 μm to 0.1 cm.

8.3 RF SENSING

Radio frequency techniques are also used in wireless sensors. RF sensors are a noncontact and intrinsically safe way to measure velocity and distance, detect motion and pressure, and indicate direction of motion. In addition, RF sensors are used for liquid-level sensing and detecting the presence of foreign objects. RF sensing techniques include: surface acoustic wave (SAW) sensors, Doppler radar, sonar, ultrasonics, and microwave sensors. Some of these sensing techniques are used in vehicle antitheft and remote entry systems.

8.3.1 SAW

Surface acoustic wave technology is being used for both RF communication and sensing. SAW devices are used to enhance the stability of oscillators in transmitters and as front-end filters to set the bandwidth and improve intermodulation performance of receivers. SAW devices respond to temperature, pressure, force, and vibration. The SAW device placed in an oscillator circuit, as shown in Figure 8.5, provides a frequency variation that can

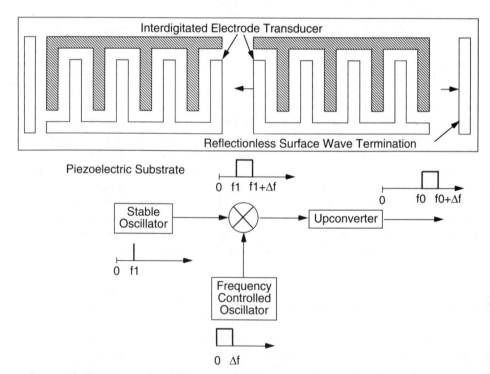

Figure 8.5 SAW device.

accurately be measured [11]. Differential techniques can be used to compensate for unwanted stimuli. The frequency range can be from dc to several GHz.

SAW devices are also being used in chemical sensors. In gas sensors, dual SAW oscillators are frequently used. One SAW device acts as the reference and the other has a gas-sensitive film deposited between the input and output interdigitated transducers. The relative change in the SAW oscillator frequency is directly proportional to gas concentration. In one study, porous oxide coatings were used as the discriminating element in the sensor [12]. The coating microstructure was evaluated by monitoring nitrogen adsorption-desorption at 77K using SAW devices as sensitive microbalances. The frequency response of a SAW device coated with zeolite in a sol-gel microcomposite film was measured at -6,350 Hz for methnol, -10,200 Hz for isopropanol, and +74 Hz for iso-octane.

Automotive navigation systems are also potential users of SAW devices. The SAW device is used in an RF-ID tag to identify a vehicle for billing purposes. In this application, SAW devices operate at 856 MHz (see Section 8.3.7).

8.3.2 Radar

Radar (radio detecting and ranging) uses reflected radio waves to measure range, bearing, and other parameters. A recently developed radar proximity sensor (shown in Figure 8.6)

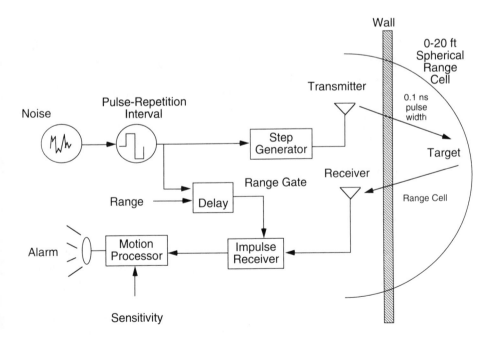

Figure 8.6 Radar sensor schematic.

uses spread-spectrum techniques and is a good example of critical elements of a radar system [13]. This particular sensor is designed to meet cost-sensitive applications such as blindspot detection in automobiles. The approach detects the echoes that reflect back over a limited range from continually propagating electromagnetic impulses. The sensor is not affected by most materials, so its location is not a factor in its ability to function properly. The unit can see through walls and distinguish between the materials of a wood or metal wall. The radar noise is coded, so an unlimited number of sensors can be colocated without interfering with each other.

The proximity sensor's antenna transmits a subnanosecond pulse at a noise dithered repetition rate. The pulse can be from 50 ps to 50 ns long. Dithering randomizes the time of the transmission in a spread spectrum that appears as noise to other detectors. The echoes returned to the (patented) receiver are sampled typically at a frequency of 1 MHz and in a timeframe (plus delay) determined by the transmitter. The range of this sensor is limited to 20 ft (6m) or less. However, several applications are possible with the sensor, including car security, voice-activated navigation, and detectors for locating one material embedded in another, such as steel within concrete.

Production microwave proximity sensors operate on radar principles. A simple microwave sensor consists of a transceiver, antenna, and signal-processing circuitry [14]. The microwave frequency range and its various subclassifications are indicated in Table 8.1.

The transceiver in a microwave sensor uses a Gunn diode or field effect transistor (FET) to generate a low-power microwave signal. The energy developed is focused by the antenna into a beam whose size is determined by the application. The reflection of the beam from an object provides a lower level signal that can be analyzed by the signal-processing circuitry. The signal can be used for Doppler shift (motion, speed, direction), strength (presence), or phase change (distance), depending on the application.

Microwave sensors are experiencing size reductions and performance improvements from the integration possible in microwave monolithic integrated circuit (MMIC) chips. A MMIC can have all or large portions of the circuit on a single chip. Flip-chip packaging (see Chapter 10) of a MMIC containing the detector diodes has allowed frequency operation up to 40 GHz [15].

8.3.3 Global Positioning System

The global positioning system (GPS) is based on information supplied by 24 satellites that are located in six orbital planes [16]. The satellites pass over the Earth at an altitude of 20,183 km (10,898 nautical miles). A single satellite will orbit the Earth twice for each Earth rotation, tracing exactly the same path twice each day, but passing four minutes earlier than the day before. The design of the system ensures that at least four satellites are in view at any one time, on or above the surface of the Earth under all weather conditions.

GPS satellites transmit at two frequencies: L1, centered at 1,575.42 MHz, and L2, centered at 1,227.6 MHz. Each satellite broadcasts a navigation message that includes a description of the satellite's position as a function of time, an almanac, and clock correction terms. Each message is comprised of 25 frames, each 30 seconds long. Commercial GPS systems are capable of locating to within 100m.

GPS technology is an integral part of the automotive intelligent transportation system (ITS) (see Section 8.3.6). GPS can be used to track field personnel and for locating position relative to location on a CD-ROM map. GPS products have been developed for consumer use. In addition to the RF components necessary to convert the signals from satellites, a 32-bit microprocessor is also required to compute the position algorithm quickly and provide frequent updates to the user. This complex sensor (receiver) is contained within a board that is 50.8 by 82.6 by 16.3 mm [17]. Increased acceptance and application in high-volume systems such as the ITS (see Section 8.3.6) will allow further component reduction through increased integration.

8.3.4 Remote Emissions Sensing

The remote measurement of vehicle's exhaust emissions is one way that remote sensing will start to affect a broad range of the population. Identifying high CO (carbon monoxide) and HC (hydrocarbon) emitting vehicles on city streets is possibly the next step toward reduced emissions from vehicles with combustion engines. The remote sensing system is shown in Figure 8.7 [18]. The main parts of the system include the infrared (IR) detector and source, a video camera to record the license plate number, a modified police radar gun, and a personal computer with specially developed software.

The remote sensing device (RSD) system operates by continuously monitoring the

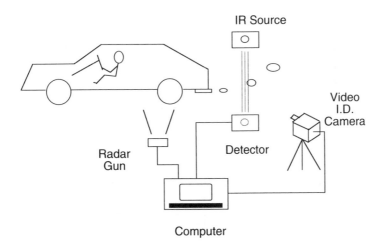

Figure 8.7   Remote measurement of vehicle exhaust emissions.

intensity of an IR source. The presence of a vehicle is indicated when the beam is broken, resulting in the reference voltage dropping to zero and a span voltage measurement being made. The value prior to beam interruption is also stored. As the vehicle exits the beam, samples are taken at 125 Hz for over 1 sec. The CO_2 (carbon dioxide) spectral region is filtered at 4.3 μm and the CO region at 4.6 μm to isolate these values. The HC values can be detected as well. This system incorporates three noncontact remote measurement devices, including an extension of previously used police radar technology that typically has a maximum range of 800m.

8.3.5 Remote Keyless Entry

Available wireless remote control of the door locks on vehicles is a step towards an automatic driver sensor (ADS). The ADS will be able to identify that a specific driver is approaching the vehicle and that the driver will require access and possibly different control settings than the previous driver. Today's remote entry systems have either IR or RF sensors. One RF system, diagrammed in Figure 8.8, has a transmitter in the ignition key that is powered by a lithium battery [19]. The transmitter in this design generates a frequency of about 41 MHz when a switch in the key is depressed. The vehicle's rear window defogger acts as the antenna. The antenna receives the signal, which is then sent to the receiver for amplification, FM detection, and wave modification. If the transmitted

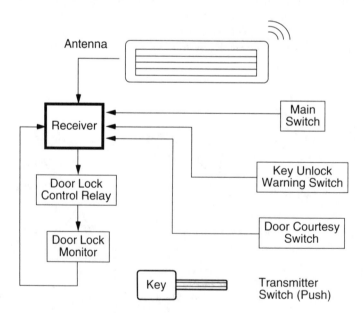

Figure 8.8   Remote door-lock control system.

code matches the stored code, the motor drive circuit either unlocks or locks the doors. The wireless system can be activated within 3 ft of the vehicle. It has a standby mode that is intermittent to avoid excessive drain on the vehicle's battery. If 10 or more incorrect codes are transmitted to the receiver within a 10-minute period, the system reacts as if a theft was occurring and all reception is discontinued. The key must be inserted manually to verify ownership.

8.3.6 Intelligent Transportation System

The intelligent transportation system (ITS), formerly known as intelligent vehicle-highway system (IVHS) in the United States and by other names (and nicknames, including smart cars and smart highways) around the world, requires several sensing technologies, many of which are RF-based, to implement full functionality. ITS has defined several areas that are being addressed to varying degrees, from research to production systems. These areas attempt to solve a number of system-related problems as well as provide new functions and services to drivers. The application areas include advanced traffic management systems (ATMS), advanced traveler information systems (ATIS), advanced public transportation systems (APTS), advanced vehicle control systems (AVCS), advanced rural transportation systems (ARTS), and commercial vehicle operations (CVO).

The various systems require a number of sensors including global positioning satellite (GPS) sensing, closed-circuit TV monitoring, IR detectors, dead reckoning, automatic vehicle location, and identification [20]. Dead reckoning can be accomplished by differential wheel-speed sensing and a fluxgate magnetic compass, or a combination of inclinometers, gyroscopes, inverse Loran, electronic odometers, and/or antilock brake system (ABS) wheel-speed sensors. In addition, the cellular infrastructure and/or radio beacons play an integral part in the vehicle navigation requirements of ATMS, ATIS, APTS, ARTS and CVO. The AVCS system will require sensors to determine distance between vehicles for smart cruise control (automatic vehicle spacing) and closing distance (time to impact) for collision avoidance systems. The RF sensors and their typical frequency ranges are indicated in Figure 8.9.

One configuration of a vehicle navigation and warning (VNAW) system is shown in Figure 8.10. The smart inertial navigation system (SNS) contains the computing that provides filter and integration for the GPS input as well as the accelerometer and gyroscope inputs [21]. The output of the SNS is position information for the VNAW system.

Collision avoidance in an automotive ITS uses a 75-GHz three-beam radar unit to provide a time-to-impact warning. One system can track up to 12 objects while measuring range angle and relative velocity [22]. The radar system also employs a video camera mounted near the inside mirror to provide a 30-deg field of view. The camera detects lane markings, road edges, and objects in the vehicle's path.

Systems that operate over a shorter distance or near obstacle detection systems (NODS) for parking and blindspot detection use Doppler radar operating at 10.5 to 24 GHz to sense objects within a few inches or feet from a vehicle. Sonar has also been developed

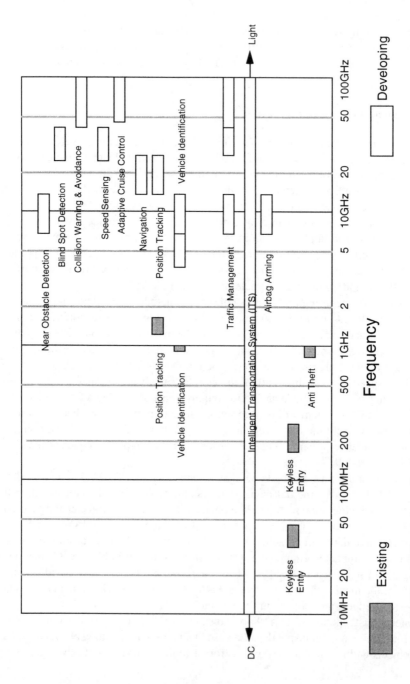

Figure 8.9 RF automotive sensing applications versus frequency.

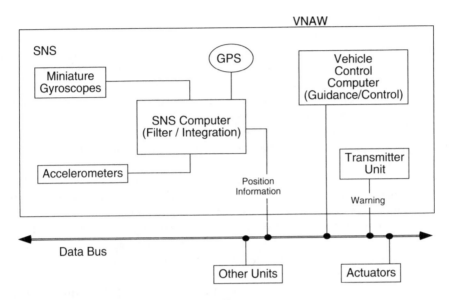

Figure 8.10 Vehicle navigation and warning (VNAW) system.

for these shorter range measurements. Sonar units are mounted at each corner of the vehicle. One sonar device operates at 50 kHz and drives a 300V transducer to generate the sonic pulse.

True road speed is one of the inputs that is required in the vehicle navigation portion of ITS. A microwave Doppler radar sensor-mounted at the front of the vehicle can provide this measurement. Intrinsic errors due to the spread of the Doppler frequencies from the top to the bottom of the beam, sensor location, false signals from clutter, pitching, vibration, signal glint (irregular ground reflections), short-term oscillator instability, and dirt are among the error possibilities for this speed-sensing approach [23]. However, microwave sensors are not significantly affected by humidity, temperature, and air movement and do not require isolation of transmitter and receiver, which are considerations for ultrasonic Doppler sensing.

The complete ITS contains a number of other systems that ultimately accomplish a number of goals, including safer travel on urban and rural roads and higher usage of the existing highway system. Certain systems and subsystems can and are being implemented separately. One element of ITS already achieving high volume and usage in a variety of other applications is RF-ID tags.

8.3.7 RF-ID

RF identification (RF-ID) tags are being used to track inventory in warehouses; work-in-process (WIP) in manufacturing plants; and animals in laboratories, on farms, or in the wild;

as well as for automatic toll collection for vehicles. The RF tag is a transponder that is read and decoded by an RF reader. For inventory and WIP in harsh environments or non-LOS operations, RF-ID is an alternative to a bar-code system [24]. The tags can be active or passive. Active tags have onboard batteries and larger data capacities, approaching 1 Mb. Decision-making ability is included in some varieties with the addition of a microcontroller. Passive tags obtain their operating power from RF energy transmitted from an antenna and as a result have limited data capacities, typically 1 to 128 bits.

RF-ID systems operate in both low and high-frequency ranges. Low-frequency systems typically operate in the frequency range of 100 kHz to 1.5 MHz and also have lower data transfer rates. They are cost-effective in access control and asset tracking applications. High-frequency systems operate in the spread-spectrum range of 900 MHz to 2.4 GHz and have higher data transfer rates. The cost of high-frequency systems is more than low-frequency systems, but they can operate at distances up to 100 ft (30m).

The main blocks of a passive RF-ID tag are shown in Figure 8.11 [25]. An RF burst is received, rectified, and used to charge a capacitor that serves as the power supply for the tag circuitry. A voltage regulator keeps the voltage across the capacitor at a constant 2V. Logic circuitry interprets the command from the RF burst. An interrogation request is answered with transmission of a number from the EEPROM. Information in the EEPROM can be changed by a transmission that enables the logic to store new data into the onboard memory. The charge pump increases the 2V to 15V to reprogram the EEPROM.

RF-ID tags are being used for automatic toll collection. The driver does not have to slow down for the system to detect the vehicle's ID and debit the driver/owner's account. This can reduce congestion on highways and still collect revenue to cover road construction and maintenance.

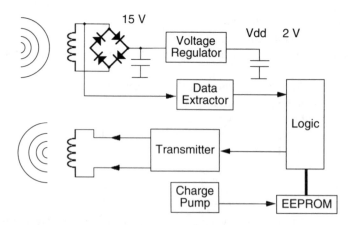

Figure 8.11   RF-ID block diagram.

8.3.8 Other Remote Sensing

RF bar-code readers demonstrate the freedom of operating untethered for a commercial application, which has become commonplace. Available RF units provide data from charge-coupled devices (CCD), laser scanners, or infrared wand sensors to a base station that can be up to 150 ft (45.7m) away. The units are mobile and can be used without the interference of wires in high-traffic areas.

Remote meter reading is being accomplished by transmitting the amount of gas, water, or electricity consumed at a particular site to a mobile receiving unit. A single operator can verify the status of almost twice the number of accounts, thereby reducing the cost of the measurement. Remote bar-code readers and scanners are also among the applications that are taking advantage of RF technology in a sensing environment.

8.3.9 Measuring RF Signal Strength

Exposure, especially continued exposure, is a concern for RF signals. The American National Standards Institute (ANSI) has limits for equivalent permissible exposure, depending on the frequency. An RF dosimeter has been designed to detect and record the strength and duration of electric fields present in work areas of naval vessels [26]. The potential to include the electronics in this already pocket-sized sensor with other RF measurement techniques may provide useful data in future applications.

8.4 TELEMETRY

Telemetry is a remote measurement technique that permits data to be interpreted at a distance from the primary detector. Telemetry is used in race cars to allow the pit crew to analyze the real-time data generated from a vehicle on the track and provide feedback to the driver that can affect the outcome of the race. For example, in Ford's Formula One system, more than a dozen sensors placed on the vehicle in various places provide information to the engine control computer and transmit the data to a mobile laboratory that travels to these races [27]. An indication of a potential problem can initiate a pit stop before the problem becomes a race-disqualifying failure. Indianapolis 500 and National Association of Stock Car Auto Racing (NASCAR) vehicles also use similar telemetry. In fact, proving ground vehicles equipped with telemetry systems allow automotive engineers to evaluate development vehicles on a high-speed test track from their offices. Cellular communications with the driver provides on-the-spot direction for performing various tests with real-time feedback on how the system is affected. In both racing and vehicle development, time is short and cost investment is high. Telemetry makes the outcome more predictable.

Telemetry is also used on production vehicles in systems that monitor pressure and temperature for each tire. The RF transmitter at each wheel sends a signal (355 or 433.92

MHz are common frequencies) that is received by a unit mounted under the dash, and a dashboard-mounted display provides information to the driver. One system uses PCM to transmit digital data by turning the carrier frequency on or off, producing burst of RF energy [28]. Switching rate and time are controlled to create the code.

Wheel-mounted transmitters deliver 10 frames of coded data in 128-ms bursts at approximately 30- to 35-sec intervals. Each frame has 8 bits of data and a blanking period. Two bits indicate tire pressure, two identify the wheel being measured, and the other four identify the car model and year. Tire pressure and temperature are important vehicle measurements that not only affect the performance and economy of the vehicle, but also have a significant impact on vehicle safety. These systems may be common on future vehicles due to the combined capability and continued cost reduction of integrated sensing and RF technology.

Another "difficult to get at" measurement is the temperature of the piston in an internal combustion engine. Telemetry has been used to transmit the temperature indicated by the variation of a temperature-sensitive chip capacitor. The system schematic is shown in Figure 8.12 [29]. Increasing temperature decreases the capacitance, which in turn increases the transmitting frequency. A multiloop antenna in the oil pan receives the transmitted signal. Data is converted to temperature through a calibration curve. The system has seven data channels to map temperatures at a number of locations inside the piston.

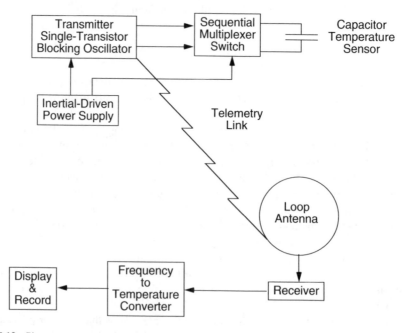

Figure 8.12 Piston temperature telemetry.

A small 22-pin hybrid package contains the multiplexer and the blocking oscillator. Power is generated based on the piston's movement. A minimum engine speed of 1,200 rpm is necessary, and temperatures from 150 to over 600°F (65 to 315°C) can be measured. To keep track of the data being sent, a reference capacitor that is not temperature-dependent transmits a 1-MHz signal. The remaining channels are each turned on for 1 sec and off for 1 sec. Temperature-dependent frequencies are generated between 250 kHz to 500 kHz for the data channels.

Ambulatory monitoring of critically ill patients or patients requiring real-time diagnostics for analysis purposes has increased the use of medical telemetry. This telemetry can be short range (within a hospital floor) or long range (for an entire wing). Most telemetry is within the 174–216 MHz (VHF) frequency band for ultralow power transmission. Higher power requirements use an alternative UHF band (450–470 MHz) and require licensing.

Real-time monitoring of noninvasive blood pressure, partial pressure of oxygen (SpO2), and peripheral pulse provides additional diagnostic and assessment information on many cardiac and respiratory patients. Electrocardiograms (ECG) and SpO2 can be monitored by two cigarette-pack-sized units weighing only a total of 380 grams [30]. A central station receives the data, analyzes the data for anomalies, and routes waveform data and parameter data to additional analysis or recording equipment. Consolidating several patients into one remote station can lower hospital costs by reducing the staff for intensive care and postoperative units, as well as provide an improved level of patient care. More data can be accumulated and analyzed with less effort from nurses and doctors. In addition, the onset of problems can be detected quickly and appropriate action can be taken sooner.

RF telemetry has also been investigated for microminiature transducers for biomedical applications. Transferring data and power into and out of the body to implanted transducers is a critical area due to the reliability of the component and, more importantly, the restrictions and potential of infection to the patient. A microstimulator was developed that measured only 1.8 by 1.8 by 9 mm^3 using RF telemetry operating at 1 MHz for power and control [31]. The assembly includes a micromachined silicon substrate that has the stimulating electrodes, CMOS, and bipolar power regulation circuitry; a custom-made glass capsule electrostatically bonded onto the silicon carrier to provide a hermetically sealed package; hybrid chip capacitors; and the receiving antenna coil. The application of microtelemetry will be more practical with improvements in integrated micromachining and reduction in the receiving antenna size.

8.5 SUMMARY

Real-time data acquisition systems will increasingly look to RF communications for faster installation and easier maintenance. A variety of wireless data services is being developed that the savvy systems developer can use to create a new sensing systems. Field service workers and other mobile data collectors who need to communicate the results of a sensed parameter or acquired data will not want to be connected by wires to a distributed system. Home, office, or industrial monitoring systems are easier to implement using RF signal transmission than using hard-wired installations, even with simple two-wire systems. Also, several hazardous monitoring situations require sensing and transmitting of data to minimize exposure of humans to toxic material. These and other nonintrusive measurements are part of an exciting new area for smart sensors.

MOSAIC, Ardis, Altair, and Oncore are trademarks of Motorola, Inc.

REFERENCES

[1] Frank, R., "Improved Portable Communications through Low Voltage Silicon Design," *Portable by Design Conference*, Santa Clara, CA, Feb. 14–18, 1994, pp. PC23–PC34.

[2] Vilbrandt, P., "Wireless Data Communications," *Sensors*, May 1993, pp. 19–21.

[3] Gaston, D., "Spread Spectrum Systems: Evaluating Performance Criteria for Your Application," *Proc. of the Second Annual Wireless Symposium*, Santa Clara, CA, Feb. 15–18, 1994, pp. 489–507.

[4] 1994 Motorola Communications Resource Guide, BR1444/D, Motorola, Inc.

[5] Leonard, M., "Digital Domain Invades Cellular Communications," *Electronic Design*, Sept. 17, 1992, pp. 40–52.

[6] Phillips, B., "Pumping Data into Cellular," *OEM Magazine*, Sept. 1994, pp. 32–41.

[7] Mathews, D. J., and C. L. Fullerton, "Microwave Local Area Network for the Computer Office," *Applied Microwave*, Winter 91/92, pp. 40–50.

[8] Nass, R., "Error-Free Wireless Data Transmission Can be Embedded in a PCMCIA Card," *Electronic Design*, Dec. 5, 1994, p. 48.

[9] Alexander, J., R. Aldridge, and D. O'Sullivan, "Wireless Zone Sensors," *Heating/Piping/Air Conditioning*, May 1993, pp. 37–39.

[10] "Wireless Data Loggers Help Pinpoint Cleanroom Humidity Fluctuations," *Microcontamination*, Aug. 1994, p. 36.

[11] Elachi, C., *Spaceborne Radar Remote Sensing: Applications and Techniques*, New York, NY: IEEE Press, 1988.

[12] Frye, G. C. et al., "Controlled Microstructure Oxide Coatings for Chemical Sensors," *IEEE Solid-State Sensor and Actuator Workshop*, Hilton Head, SC, June 4–7, 1990, pp. 61–64.

[13] Ajluni, C., "Low-Cost Wideband Spread-Spectrum Device Promises to Revolutionize Radar Proximity Sensors," *Electronic Design*, July 25, 1994, pp. 35–38.

[14] "Microwave Sensors," *Measurements & Control*, December, 1992, p. 173.

[15] Fischer, M. C., M. J. Schoessow, and P. Tong, "GaAs Technology in Sensor and Baseband Design," *Hewlett-Packard Journal*, April 1992, pp. 90–94.

[16] Harris, C. and R. Sikorski, "GPS Technology and Opportunities," *Expo COMM China '92*, Beijing, China, Oct. 30–Nov. 4, 1992.

[17] VP Oncore GPS Receiver, Motorola Brochure, 1994.

[18] Glover, E. L., and W. B. Clemmens, "Identifying Excess Emitters with Remote Sensing Device: A Preliminary Analysis," SAE 911672, Warrendale, PA.

[19] McCarty, L. H., "Coded Radio Signal Locks/Unlocks Car's Doors," *Design News*, Jan. 22,1990, pp. 108–109.

[20] Sweeney, Jr., L. E., "An Overview of IVHS Sensor Requirements," *Proc. of Sensors Expo West*, San Jose, CA, March 2–4, 1993, pp. 229–233.

[21] Maseeh, F., "Microsensor-Based Navigation and Warning Systems: Applications in IVHS," *Proc. of Sensors Expo West*, San Jose, CA, March 2–4, 1993, pp. 251–255.

[22] Sawyer, C. A., "Collision Avoidance," *Automotive Industries*, Jan. 1993, p. 53.

[23] Kidd, S. et al., "Speed Over Ground Measurement," SAE Technical Paper 910272, Warrendale, PA.

[24] Rishi, G., "RF Tags in Manufacturing," *ID Systems*, Nov. 1994, pp. 51–54.

[25] McLeod, J., "RF-ID: A New Market Poised for Explosive Growth," *Electronics*, Feb. 8, 1993, p. 4.

[26] Rochelle, R. W. et al., "A Personal Radio-Frequency Dosimeter with Cumulative-Dose Recording Capabilities," *Proc. of Sensors Expo 1990*, Chicago, IL, Sept. 11–13, 1990, pp. 107B-2–107B-9.

[27] "Telemetry: Racing into Your Future," Ford Electronics Brochure, Dearborn, MI.

[28] Siuru, Jr., W. D., "Sensing Tire Pressure on the Move," *Sensors*, July 1990, pp. 16–19.

[29] Murray, C. J., "Telemetry System Monitors Piston Temperatures," *Design News*, Oct. 2, 1989, pp. 192–193.

[30] "Medical Telemetry To Wireless," *Medical Electronics*, Oct. 1993, pp. 106–107.

[31] Akin, T. et al., "RF Telemetry Powering and Control of Hermetically Sealed Integrated Sensors and Actuators," *IEEE Solid-State Sensor and Actuator Workshop*, Hilton Head, SC, June 4–7, 1990, pp. 145–148.

Chapter 9

Microelectromechanical Systems (MEMS)

"And these other instruments, the use of which I cannot guess?"

"Here, Professor, I ought to give you some explanations. Will you be kind enough to listen to me?"

—Jules Verne, *Twenty Thousand Leagues Under the Sea*

9.1 INTRODUCTION

The micromachining technology that has enabled semiconductor sensors is being applied to control systems for numerous mechanical applications. In some cases, the microstructures that are being developed have no direct relationship to sensors. However, many of these devices enhance the performance of the total system or allow system design that was not previously possible. Since micromachining is fundamental to manufacturing these structures, future developments in this area will improve both micromachining technology and the systems that utilize them.

The micro-level design and fabrication of mechanical structures is called microelectromechanical systems (MEMS). The extent to which micromachining is utilized to produce components that are not sensors is demonstrated by the list of developed silicon structures shown in Table 9.1 [1]. The batch processing of mechanical components has the same potential for mechanical engineering and other disciplines that semiconductor batch-processing has had for electrical engineering. Today's $100-plus billion semiconductor industry and resulting electronics industry would not exist without batch-processing technology. A few of the areas that are being explored will be discussed in this chapter to indicate the variety and extent of MEMS technology.

9.2 MICROMACHINED ACTUATORS

Actuators micromachined from silicon and other semiconductor materials use electrostatic, electrothermal, thermopneumatic, electromagnetic, electro-osmotic, electrohydrodynamic,

Table 9.1 Micromechanical Structures in Silicon

Cryogenic microconnectors	Microprobes
Fiber optic couplers	Micropumps
Film stress measurement	Microswitches
Fluidic components	Microvacuum tubes
IC heatsinks	Microvalves
Ink jet nozzles	Nerve regenerators
Laser beam deflectors	Photolithography masks
Laser resonators	Pressure switches
Light modulators	Pressure regulators
Membranes	RMS converters
Microaligners	Thermal print heads
Microbalances	Thermopile
Microfuses	Torsion mirrors
Microgears	Vibrating microstructures
Micromolds	Wind tunnel skin friction
Micromotors	
Micropositioners	
Micro-interconnects	
Microchannels	
Microrobots	
Micromanipulator/handler	
Micromechanical memory	
Microchromatograph	
Microinterferometer	
Microspectroscopy	
Micro scanning electron microscope	

After: [1].

and other means to provide motion. These actuators range from existing products that provide unmatched performance compared to their macro-scale counterparts to intriguing lab curiosities that require significant development to become practical. Examples of a silicon microvalve, micromotor, micropump, and microdynamometer as well as actuators in alternate materials will be discussed in this section.

9.2.1 Microvalves

One area of MEMS that has achieved production status is microvalves. Figure 9.1 shows one design, a silicon Fluistor™ (or fluidic transistor) microvalve that is approximately 5.5 mm by 6.5 mm by 2 mm [2]. The bulk-micromachined cavity in the top section is filled with a control liquid. In the unactivated state, gases can flow through the valve. A voltage applied to the heating element on the diaphragm causes sufficient expansion of the liquid to deflect the diaphragm, close the valve seat, and restrict flow. The valve has a dynamic

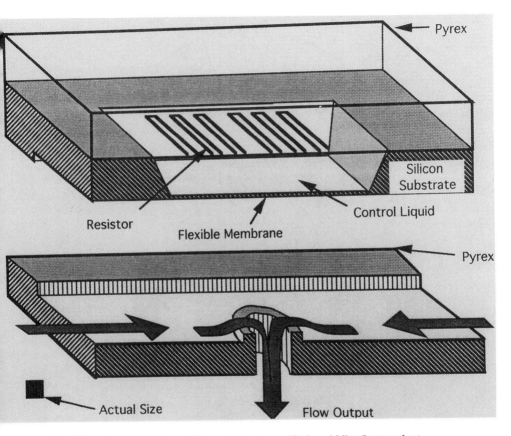

Figure 9.1 Microvalve construction and flow path (*courtesy of* Redwood MicroSystems, Inc.).

range of 100,000 to 1, controlling gas flows from 4 microliters per minute to 4 liters per minute at a pressure of 20 psi. A normally closed design is also available.

9.2.2 Micromotors

Micromotors are among the more interesting demonstrations of the future potential of MEMS. Electrostatic and electromagnetic motors have been fabricated by several researchers. An example of an electrostatic motor is shown in Figure 9.2 [3]. The rotor's diameter is typically about 0.1 mm. The stator is activated by pulses that produce an electrostatic force. Coupling the motor to a load and friction are among the problems that must be solved to make the motor useful. However, a fan for cooling ultrahigh-performance microprocessors is among the possibilities that can be envisioned for this technology.

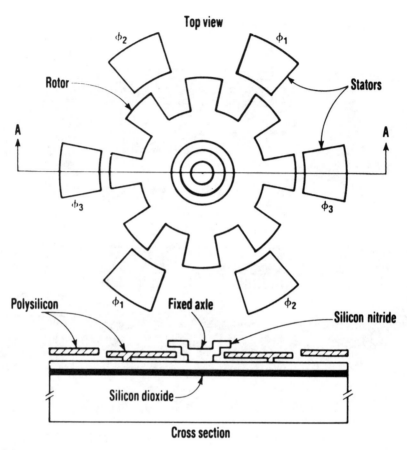

Figure 9.2 Electrostatic micromotor (*courtesy of Electronic Design* [3]).

The coil windings in a magnetic micromotor require a thicker actuator cross-section. One approach uses a polyimide-based process that allows microstructures to be fabricated on top of a standard CMOS process [4]. The core of the motor is multilevel electroplated nickel-iron wrapped around a meander conductor. The meander structure for conductor and coil is accomplished by reversing the normal roles of the conductor and magnetic core. The magnetic core is wrapped around a planar conductor, either by interlacing or by interconnecting multilevel metal layers. Polyimide is used as the dielectric interlayer between for embedding the coils and the cores. Figure 9.5 shows the functional micromotor and an example of the meander coil [4, 5]. The rotor was operated up to a speed of 500 rpm based on the limitations of the drive controller.

(a)

Figure 9.3 (a) Magnetic micromotor, and (b) multilevel meander coil (*courtesy of* Georgia Institute of Technology).

9.2.3 Micropumps

A miniature peristaltic pump has been designed, fabricated, and tested. The pump consists of three silicon wafers bonded together as shown in Figure 9.4 to produce a flow channel, membrane, and heater [6]. Silicon fusion bonding was used to bond the wafers containing these elements together. The heaters are suspended in a thermopneumatic fluid and sequentially activated from left to right. The flow channel is fabricated by a proprietary etching technique to closely match the contour of the bulging silicon nitride membrane. Heating the fluid deflects the membrane that displaces the liquid. The deflected membrane seals the channel to prevent backflow. Performance of the micropump has been predicted to be 7 microliters per minute at 15 psi. Pumps that can displace precise amounts of liquid have applications in medicine for automatic insulin dispensing as well as in manufacturing for precise process control and reduced material cost.

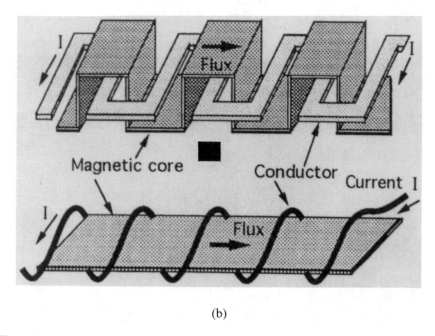

(b)

Figure 9.3 Continued.

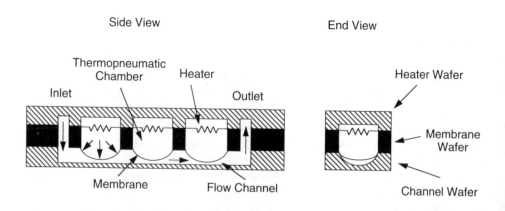

Figure 9.4 Three thermopneumatic actuators provide peristaltic pumping action.

9.2.4 Microdynamometer

The first steps have been taken to achieve a functional planar microdynamometer [7]. A dynamometer consists of a motor, a coupling gear train, a generator to act as an active load, and associated electronics. The SLIGA process (LIGA process with the addition of a sacrificial layer discussed in Section 2.4.1) was used to fabricate mechanical components of the microdynamometer. Magnetic actuation was chosen for the micromotor and generator windings. As shown in Figure 9.5, a 2-pole-pair motor with three windings per pole (upper right hand corner) and a generator with six windings were fabricated using electroplated nickel. Photodiodes were integrated in the design that can be used to determine the position of the rotor in the motor and the generator. Among the issues that must be resolved to realize a functional microdynamometer are magnetic materials problems.

Figure 9.5 Microdynamometer fabricated with electroplated nickel. A single-pole-pair test drive is at the bottom right (*courtesy of* the University of Wisconsin).

9.2.5 Actuators in Other Semiconductor Materials

Actuators have been designed and fabricated using thin film and plated metals, dielectrics, and photoresists for sacrificial layers on gallium arsenide (GaAs) and indium phosphide (InP) substrates. GaAs and InP materials are used to fabricate monolithic microwave integrated circuit (MMIC) devices. MMIC devices require tuning that potentially can be performed by on-chip actuators and result in improved performance and yield. Sliding interdigitated capacitive tuners, bending beams, and rotating switches have been fabricated in a MMIC-compatible process as a first step toward their ultimate use to control the MMIC [8].

Epitaxial 3C-SiC (silicon carbide) films and sputtered amorphous SiC films have been investigated for high-temperature MEMS devices [9]. Suspended diaphragms and freestanding cantilever structures were etched into epitaxial films using bulk micromachining. The cantilever beams deflected downward due to residual stress variation in the film. Surface micromachining techniques were also applied to amorphous SiC. A 150-μm diameter, 1.5-μm thick gear was fabricated from amorphous SiC sputtered on silicon dioxide. These materials hold promise for MEMS devices, but require considerable development effort to progress from the laboratory into production.

9.3 OTHER MICROMACHINED STRUCTURES

In addition to actuators, MEMS structures will be used for system components that require small size or reproducibility that can be achieved with micromachining. (Microgears were already discussed in the microdynamometer section.) Multiple metal microgears were driven by forced air or relatively weak magnetic fields. Other examples will be discussed in this section that demonstrate the variety of research and development activity that is occurring for MEMS-based devices.

9.3.1 Cooling Channels

A micro heat pipe created by parallel microchannels in the bottom of a high-performance IC can provide cooling to minimize hot spots, improve performance, and increase reliability. A proposed approach is shown in Figure 9.6 [10]. After etching the channels, multiple metalization layers are vacuum deposited to line the walls of the channel and seal the top. Heating the chip in a fluid bath fills approximately 20% of the cavity volume with fluid. The ends are sealed after this step to contain the fluid. The micro heat pipe's operation causes fluid to evaporate in high-temperature regions and condense in low-temperature areas, resulting in a more uniform temperature distribution across the IC. Increasingly higher operating frequencies for higher performance MPUs and MCUs may require this type of cooling to avoid increasing the packaging size and/or increasing the amount of external heatsinking.

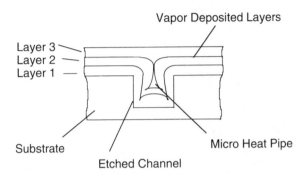

Figure 9.6   A micro heat pipe for cooling integrated circuits.

9.3.2 Micro-Optics

A number of micro-optical devices have been produced by micromachining techniques, including gratings, lenses, air bridges, electrical interconnects, fiber-optic couplers, alignment aids, corner reflectors, and waveguides [11]. In addition to silicon, a number of III-V semiconductor materials are being investigated for improved optical performance. A chemically assisted ion beam etching (CAIBE) process using argon and xenon ion beams and chlorine as the reactive gas was used to fabricate surface-emitting lasers. The 3-μm diameter lasers operated at room temperature with a threshold current below 1.5 mA and a differential quantum efficiency of 16%. However, a number of problems must be solved for the lasers to provide a viable solution for optical computing, chip-to-chip communications, and optical switching. The combination of mechanics, optics, and electronics at the micron scale promises to be an important field in optical microsystems.

A different combination of micromachining and optics is demonstrated in Figure 9.7 [12]. This polysilicon microscanner consists of a hollow nickel-plated polygon reflector on the rotor of an electrostatic drive micromotor. Micro-optomechanical systems (MOMS) are based on wafer-level integration of optical and MEMS components. Microscanners and movable optical elements have been designed and fabricated using electroless plating of reflective nickel surfaces on the rotor of the micromotor. The thickness (height) of the nickel is 20 μm and the width is 10 μm. Scanners with diameters from 250 to 1,000 μm have been produced, but the larger micromotors do not operate reliably after they are released. Diffraction grating microscanners were fabricated with spatial gratings of 2 and 4 μm using a similar process. The key to MOMS devices is the ability to fabricate different optical and mechanical structures on a common substrate.

An actuator has been proposed that uses optical power to provide mechanical energy [13]. An optical actuator has the potential advantages of higher operating speed, lower power consumption, and lower thermal expansion than nonoptical approaches. A silicon cantilever reacts to a photoelectric current by relaxing, as opposed to an applied electrostatic voltage, which stress the beam further.

Figure 9.7 A rotating polygon optical microscanner made by electroless plating of nickel (*courtesy of* Case Western Reserve University).

9.3.3 Microgripper

Microgrippers or microtweezers have been demonstrated by several researchers. One research team has developed a surface-micromachined polysilicon microgripper that is activated by an electrostatic comb drive [14]. The electrostatic comb drive technique provides the force for several microactuators as well as an oscillating structure for many sensors. The two movable gripper arms are controlled by a three-element electrostatic comb as shown in Figure 9.8. The length of the drive arms L_{dr} is 400 µm and each extension arm is $L_{ext} = 100$-µm long. By using separate open and close drivers, the gripping range for a given maximum voltage is doubled. Movement of 5 µm was produced by the grippers with less than 30V applied to the comb drive.

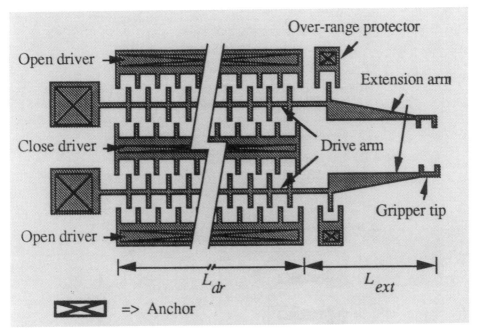

Figure 9.8 Schematic of a microgripper with an electrostatic comb drive (*courtesy of* the Berkeley Sensor and Actuator Center, University of California).

9.3.4 Microprobes

The topography of a surface can be measured by a cantilever beam contacting and scanning across the surface of a sample. This instrument, an atomic force microscope (AFM), requires a mechanical structure with a sharp tip, small spring constant, and high resonant frequency [15]. Batch fabrication yields cantilevers with very reproducible characteristics, and piezoelectric, capacitive, or piezoresistive sensing techniques can be used to sense the deflection of the probe tip. The construction of an AFM is shown in Figure 9.9. The dimensions of one device are $L1 = 175$ μm, $L2 = 75$ μm, $w = 20$ μm, $b = 90$ μm, and $t = 2$ μm. The calculated spring constant of this structure is 4 N/m. The AFM was used to measure a silicon dioxide grating that had a depth of 270 Å and repeated every 6.5 μm. These types of devices may be useful in profilometry and IC inspection.

9.3.5 Micromirrors

Digital micromirror devices have been micromachined that may be used for displays. As shown in Figure 9.10, the micromirror element is an aluminum mirror suspended over an air gap by two thin post-supported hinges [16]. The mechanically compliant hinges permit the mirror to rotate 10 deg in either direction. The posts provide the connections to a bias/reset bus that connects all the mirrors of the arrays to a bond pad. The mirrors are

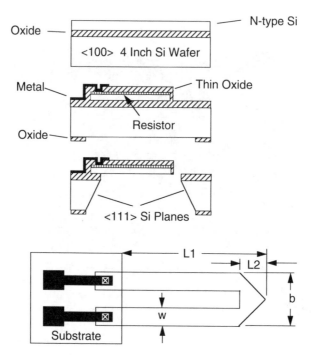

Figure 9.9 Atomic force microscope probe.

fabricated over conventional CMOS static random access memory (SRAM) cells that provide an address circuit for each mirror. The mirrors have a response time of approximately 10 ms and can be pulse-width modulated to provide a gray-scale output in a black and white display. Monolithic arrays with 768 by 576 pixels have been demonstrated.

9.3.6 Heating Elements

Multijunction thermal converters (MJTC) have been fabricated using a standard CMOS process and bulk micromachining [17]. After CMOS processing, a cavity is etched by bulk micromachining that produces a suspended MJTC cantilever structure, as shown in Figure 9.11. Polysilicon resistive heating elements and aluminum-polysilicon thermocouple junctions are encapsulated in glass. The glass protects these elements from the etchant and provides a mechanical support. MJTCs have potential applications in low-cost, high-precision RF and microwave power circuits.

9.3.7 Thermionic Emitters

Arrays of sputtered tungsten thermionic emitters have been fabricated by surface micromachining [18]. An SiO_2 layer isolates sputtered tungsten from the silicon substrate. The

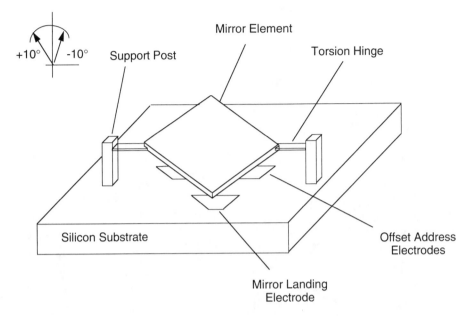

+10° -10° Support Post Mirror Element Torsion Hinge

Silicon Substrate Offset Address Electrodes

Mirror Landing Electrode

Figure 9.10 Digital micromirror with torsion hinge suspension.

tungsten is patterned by wet-etching, and the silicon is also wet-etched under the filament to avoid contact. The filaments were tested in a vacuum of 5 by 10^{-7} torr. The filaments changed from barely visible red to white light once they start to emit. Operating life approaching 1 hr has been achieved with emission currents of about 10 nA. The thermionic emitters are used as the first stage on a miniature scanning electron microscope (SEM). The SEM is less than 2 cm^3 in volume and is formed by stacking five silicon dice.

These recent developments in thermionics require examining terms that are commonly used for semiconductors. The invention and volume production of semiconductors or *solid-state* devices displaced vacuum tubes. However, thermionic devices made using semiconductor processes have now created *gaseous-state* microdevices. Volume production of these devices is certain to raise comments if they are referred to as solid-state devices.

9.3.8 Field Emission Devices

A number of field emission devices (FED) are being developed using micromachining techniques [19]. An FED consists of an array of emitting microtips. In one approach, the microtips are formed by etching chambers inside a masked silicon wafer. Molybdenum tips are vacuum deposited in the chamber. Several hundred emitter tips are fabricated for each pixel, allowing dozens of tips to fail without discernible loss of brightness.

Another technique to produce the microtips uses selective etching of a polycrystalline

Figure 9.11 A cantilever multijunction thermal converter (MJTC) (*courtesy of* NIST and Ballantine Laboratories, Inc.).

silicon substrate. This is a self-aligned process that uses the crystal structure of silicon to produce atomically sharp silicon tips. It may be possible to eliminate lithography in this approach due to the self-aligning nature of the process. The application of FEDs is discussed further in Section 11.5.2.

9.3.9 Unfoldable Microelements

Microstructures that have elastic joints have been used to make movable three-dimensional structures from planar surface-micromachined structures. Microrobots are among the possibilities for these devices. The basis of the movable structure is shown in Figure 9.12(a) [20]. Polyimide provides the flexible connection to polysilicon plates or skeleton-like structures. The phosphosilicate glass (PSG) is sacrificially etched to free the structure for the substrate. Released structures have actually been folded at the hinges like paper, creating microcubes and insect-like three-dimensional structures. A robot ant example is shown in Figure 9.12(b) [21].

Methods for actuating the artificial arms on these devices have been studied and compared to properties of nature's actuator—muscle [22]. One of the more interesting developments is an electrostatic muscle that uses the combination of a number of small force displacements to produce significant displacement [23]. Integrated force arrays produce a flexible membrane that contracts with an applied electric field. Surface micromachining

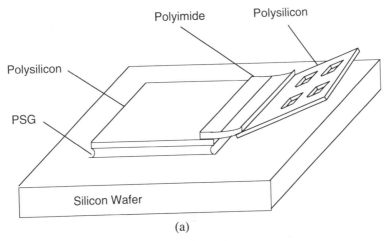

(a)

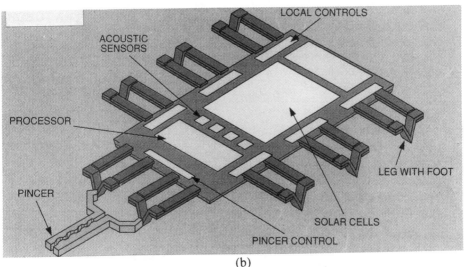

(b)

Figure 9.12 (a) Three-dimensional structure with elastic joint. (*After:* [20].) (b) Robotic ant (*courtesy of* Boston University).

is used to create metal-clad polyimide rectangles that are only a few microns on a side. An array of 1.5 million cells forms a 1-cm long fiber that contracts 0.3 cm with only a few volts applied. The force array technology is being developed for coordinated movement of interlocking flexible elements. Besides robotics, self-aligned high-density multicontact electrical connectors can be designed using this technology.

9.3.10 Micronozzles

Microminiature apertures and nozzles are required for optical instruments and a variety of mechanical devices, including high-resolution ink jet printers, flow control, and atomizers [24]. Sacrificial etching is used to produce a highly cusped nozzle-shaped structure using silicon nitride. A mold is created for the nitride by steps (a) through (d) in Figure 9.13. The nitride structure is freed from the substrate by a KOH etch and precision backsawing as shown in Figure 9.13(e). Alternatively, backmasked anisotropic etching produces the structure shown in Figure 9.13(f). A variety of other materials including silicon dioxide, boron doped silicon, polysilicon, refractory, and noble metals can be used for the nozzle.

9.3.11 Interconnects for Stacked Wafers

Wafer on wafer construction is being investigated to improve the density of integrated circuits. Micromachining with subsequent metalization can provide a technique to inter-

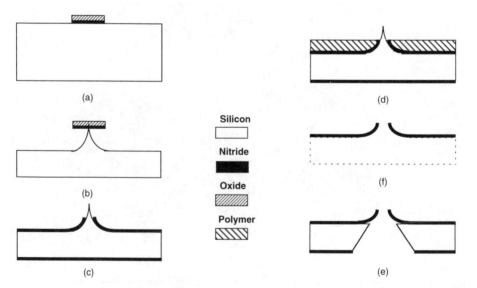

Figure 9.13   Micronozzle fabrication process.

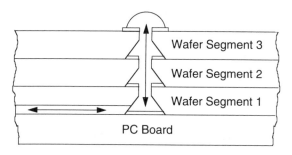

Figure 9.14   Interconnects from pyramid-shaped vias for multiple layer structure. (*After:* [25].)

connect a stack of various silicon technologies at the wafer level. As shown in Figure 9.14 [25], anisotropically etched wafers are aligned where the interconnection is required. The pyramid-shaped structure has a square opening at the top of 25 microns and is 120 microns at the bottom for thin wafers. A fine gold-plated wire mesh is used to fill the cavity. The wire mushrooms at the top and compresses at the bottom to form a gold on gold contact.

9.4 SUMMARY

Micromachining technology is used to produce micron-scale machines—movable structures—that are independently fascinating. If they can be interconnected for use in a microsystem, these devices have the potential to create new industries just as semiconductor technology has done. MEMS devices combined with sensors will provide new tools and improved performance for control systems.

Fluistor is a trademark of Redwood Microsystems, Inc.

REFERENCES

[1] Bryzek, J., and J. R. Mallon, "Silicon Integrated Circuit Sensors and Actuators," *Wescon Professional Advancement Program Session 9*, Nov. 14–15, 1989, San Francisco, CA, pp. 196–201.

[2] Zdeblick, M., "A Revolutionary Actuator for Microstructures," *Sensors*, Feb. 1993, pp. 26–33.

[3] Leonard, M., "Electric Motors on a Chip Advance from Academia's Labs," *Electronic Design*, Jan. 25, 1990, p. 26.

[4] Allen, M. G., "Polyimide-Based Processes for the Fabrication of Thick Electroplated Microstructures," *The 7th International Conference on Solid State Sensors and Actuators (Transducers '93)*, June 7–10, 1993, pp. 60–63.

[5] Ahn, C. H., and M. G. Allen, "A Fully Integrated Micromagnetic Actuator with a Multilevel Meander Magnetic Core," *Technical Digest IEEE Solid-State Sensor and Actuator Workshop*, June 22–25, 1992, Hilton Head, SC, pp. 14–18.

[6] Folta, J. A., N. F. Riley, and E. W. Hee, "Design, Fabrication and Testing of a Miniature Peristaltic Membrane Pump," *Technical Digest IEEE Solid-State Sensor and Actuator Workshop*, June 22–25, 1992, Hilton Head, SC, pp. 186–189.

[7] Christenson, T. R., H. Guckel, K. J. Skrobis, and T. S. Jung, "Preliminary Results for a Planar Micrody-

namometer," *Technical Digest IEEE*, Solid-State Sensor and Actuator Workshop, June 22–25, 1992, Hilton Head, SC, pp. 6–9.

[8] Hackett, R. H., L. E. Larson, and M. A. Melendes, "The Integration of Micro-Machine Fabrication with Electronic Device Fabrication on III-V Semiconductor Materials," *IEEE 91CH2817–5 Transducers '91*, pp. 51–54.

[9] Tong, L., M. Mehregany, and L. G. Matus, "Silicon Carbide as a Micromechanics Material," *Technical Digest IEEE Solid-State Sensor and Actuator Workshop*, June 22–25, 1992, Hilton Head, SC, pp. 198–201.

[10] Markstein, H., "Embedded Micro Heat Pipes Cool Chips," *Electronic Packaging and Production*, Oct. 1993, p. 14.

[11] Deimel, P. P., "Micromachining Processes and Structures in Micro-optics and Optoelectronics," *Journal of Micromechanics and Microengineering*, Dec. 1991, pp. 199–222.

[12] Merat, F., and M. Mehregany, "Integrated Micro-Optical-Mechanical Systems," *Proc. of SPIE*, Vol. 2383, Feb. 1995.

[13] Tabib-Azar, M., "Optically Controlled Silicon Microactuators," *Nanotechnology*, 1990, pp. 81–92.

[14] Kim, C-J., A. P. Pisano, R. S. Muller, and M. G. Lim, "Polysilicon Microgripper," *Technical Digest IEEE Solid-State Sensor and Actuator Workshop*, June 4–7, 1990, Hilton Head, SC, pp. 49–51.

[15] Tortonese, M., H. Yamada, R. C. Barret, and C. F. Quate, "Atomic Force Microscopy Using Piezoresistive Cantilever," *IEEE 91CH2817–5 from Transducers '91*, pp. 448–451.

[16] Mignardi, M. A., "Digital Micromirror Array for Projection TV," *Solid State Technology*, July, 1994, pp. 63–68.

[17] Gaitan, M., J. Kinard, and D. X. Huang, "Performance of Commercial CMOS Foundry Compatible Multijunction Thermal Converter," *The 7th International Conference on Solid State Sensors and Actuators (Transducers '93)*, June 7–10, 1993, 1012–1014.

[18] Perng, D. C., D. A. Crewe, and A. D. Feinerman, "Micromachined Thermionic Emitters," *Journal of Micromechanics and Microengineering*, Vol. 2, No. 1, March 1992, pp. 25–30.

[19] Derbyshire, K., "Beyond AMLCDs: Field Emission Displays?" *Solid State Technology*, Nov. 1994, pp. 55–65.

[20] Shimoyama, I. et al., "Insect-Like Microrobots with External Skeletons," *IEEE Control Systems*, Feb. 1993, pp. 37–41.

[21] Robinson, G. R., "Micromechanics Drives New Gears of Innovation," *Design News*, Feb. 10, 1992, pp. 23–24.

[22] Hunter, I. W., and S. Fafontaine, "A Comparison of Muscle with Artifical Actuators," *Technical Digest IEEE Solid-State Sensor and Actuator Workshop*, June 22–25, 1992, Hilton Head, SC, pp. 178–185.

[23] Brown, C., "Force Arrays Mimic Natural Motion," *Electronic Engineering Times*, June 20, 1994, pp. 41, 49.

[24] Farooqui, M. M., and A.G.R. Evans, "Microfabrication of Submicron Nozzles in Silicon Nitride," *Journal of Microelectromechanical Systems*, Vol. 1, No. 2, June 1992, pp. 86–88.

[25] Markstein, H., "Vertical Wafer Integration Optimizes Memory Density," *Electronic Packaging & Production*, Jan. 1995, p. 30.

Chapter 10
Packaging Implications of Smarter Sensors

"I regret that there are no more worlds to conquer."
—Alexander the Great

10.1 INTRODUCTION

All of the advances and research occurring in micromachining would lead one to believe that breakthroughs in these areas will be sufficient to revolutionize sensing. Unfortunately, the problems associated with the basic sensor packaging are compounded when the sensor is combined with higher levels of electronics. These problems initiate at the lowest level of die and wire bonding and extend to encapsulation, sealing, and lead forming issues. Fundamental assembly differences frequently exist between sensor and microelectronics packaging and these are among the problems that must be solved to achieve smarter sensors. These differences exist in die bonding for stress isolation instead of heat dissipation and wire-bonding procedures. Packaging is essential to establishing the reliability of the sensor. Therefore, the reliability requirements must be taken into account in the design of the package, especially for custom packages in specific applications. Testing of the sensor and circuitry combination also requires combining test capability from both technologies. This chapter will address sensor packaging technology, especially new technology from the semiconductor industry that should be applied to smart sensors and reliability concerns for sensor packages.

10.2 SEMICONDUCTOR PACKAGING APPLIED TO SENSORS

Many of today's sensor packages resemble semiconductor packages of the 1980s or even the 1970s. The semiconductor industry has made significant progress in high-density plastic-encapsulated packages. Packaging trends for the next decade are shown in Figure 10.1 [1]. The increased use of surface-mount technology (SMT) is among the more

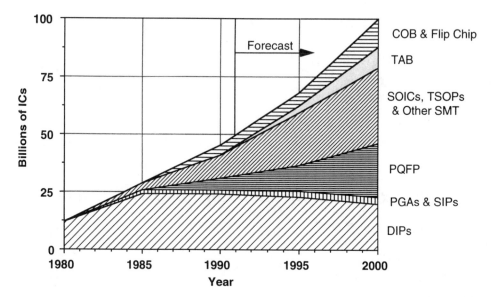

Figure 10.1 Semiconductor packaging projections. (*After*: [1].)

important observations. To achieve increased functionality without increased silicon complexity, available silicon technologies are being combined at the package level in packages based on semiconductor, not module manufacturer, assembly techniques. These multichip modules (MCM) are being evaluated for several applications, including automotive. The figure shows packaging technology acceptance for integrated circuits in general, but indicates a decline in the use of the previously popular dual in-line plastic (DIP) package. Other through-hole packages, like single in-line plastic (SIP) and pin grid array (PGA) will also not increase. New SMT approaches, like ball grid array (BGA) packages, are the focus of present packaging development and are too new to be included in the Figure 10.1. For future, highly integrated components, packaging techniques must take into account more complex, system-level requirements as well as SMT assembly requirements. The ability of the sensor industry to adapt to the newer semiconductor packaging approaches to sensors will determine the acceptance of smart-sensor technology and the future growth of the industry.

Sensor packages have basic requirements that are similar to semiconductor devices. The variety of harsh sensor applications makes packaging more difficult than packaging for a semiconductor device. However, the basic package operations occur in similar order.

A completed sensor wafer has a final processing step that prepares it for packaging [2]. This could include thinning the wafer and attachment of a backside metal such as a gold-silicon eutectic. Sensors tested at the wafer level that do not meet minimum specifications are identified as rejected units by an ink dot. Sensors are then separated into individual dice from the wafer by sawing or scribe-and-break techniques. Good sensor dice are placed in carriers that allow automatic pick and place machines to transfer dice from the carrier to the final package where a die bond attaches the sensor firmly to the package. Wire bonds connect the electrical contacts on the die surface to the leads of the package that allow the sensor to interface to external components. The package is then sealed if it is metal or ceramic, or encapsulated if it is molded plastic. Lead plating and trim (or singulation) occur next and are followed by marking and final test operations.

The actual design of the package must take into account sensitive areas of the semiconductor device and the sensor's specific function. As shown in the following list [2], the sensitivity of the semiconductor to light must be minimized in packaging for an accelerometer, but optimized for a photodiode. Similarly, the package's sensitivity to stress must be taken into account during its design to prevent this stress from affecting the offset and sensitivity in a stress-sensitive pressure sensor. For smart sensors, the key item in the list below, which shows the characteristics that affect sensor packaging, is the fact that integration level affects the sensor's package.

- Wafer thickness & wafer stack (e.g., single, silicon-silicon, silicon-glass);
- Dimensions;
- Environmental sensitivity/requirement for physical interface;
- Physical vulnerability/stress sensitivity;
- Heat generation;
- Heat sensitivity;
- Light sensitivity;
- Magnetic sensitivity;
- Integration level.

The active area of semiconductor devices is protected by a passivation layer that is deposited near the end of the wafer fabrication process. Table 10.1 [2] lists the common terms used to describe this layer. Silicon dioxide and silicon nitride doped with boron, phosphorus, or both, are two materials used for the passivation process. For the semi-

Table 10.1 Silicon Wafer Passivation Layers

Material	Assembly Level
Silox	Wafer
Vapox	Wafer
Pyrox	Wafer
Glassivation layer	Wafer
PSG (phosphosilicate glass)	Wafer
BSG (borosilicate glass)	Wafer
PBSG (phospho-borosilicate glass)	Wafer
Parylene	Package
Dimethyl-silicone	Package

After: [2].

conductor sensor, the mechanical properties of these layers must also be taken into account. For example, in silicon pressure sensors, implanted or diffused elements are protected by a passivation layer, but the diaphragm area is masked to avoid the dissimilar material interface and dampening effect that would be caused by the glass layer.

For semiconductor components, the package provides protection from environment that can include moisture and gaseous or liquid chemicals. These packages are further protected in applications like automotive underhood-mounted modules by additional epoxy and silica potting compounds or conformal coatings (e.g., acrylics, polyurethane, silicone, and ultraviolet curing compounds) that cover the printed circuit to which the component is mounted. However, semiconductor sensors frequently have to interface to this environment. Pressure sensors, for example, that respond to static and dynamic pressures must have protection techniques that allow the pressure signal to be transmitted to the force-collecting diaphragm with minimal damping and distortion of the signal. Distortion can be caused by compressible fluids. Additional protective materials for the die surface and the wire-bonding connections, such as a compliant thin conformal parylene coating or hydrostatic methyl-silicone gel, are frequently used as the means to transmit pressure to the top surface of the sensor. Parylene deposition is a vacuum process where the reactive vapor is passed over a room-temperature sensor and coats it with the polymer [3]. The equipment used to perform this process is quite sophisticated, especially relative to gel coatings. The kind of media to which parylene and gel-protected sensors can be interfaced is limited by the properties of the protective material.

10.2.1 Increased Pin Count

One of the more difficult problems that must be solved when additional electronic circuitry is integrated with or interfaced to the sensor is the requirement for additional pinouts. Integrated circuits, including microcontrollers have industry-accepted standard packages that allow a large number of pinouts. To increase the density in rapidly increasing surface-mounted applications, pin pitches below 0.5 mm are being pursued. Tape-automated bonding (TAB) is one of the technologies being developed to address fine pitch requirements. Sensors, on the other hand, are usually limited to eight or fewer pins. Furthermore, the packaging varies considerably from manufacturer to manufacturer, with no standard form factor (refer to Figure 10.2). The requirement for a mechanical interface for pressure, force, flow, or liquid level causes additional packaging problems.

Additional circuitry, whether it is simple signal conditioning or more sophisticated approaches that include microcontroller capabilities, has an impact on sensor packaging [4]. A ''sensor only'' versus integrated sensor plus control circuit is one level of differentiating packaging requirements. However, increased functionality through additional circuitry, either on the same chip or from a separate chip included in the ultimate sensor package or module, will affect the pin count, normally increasing the number of package pins. An exception occurs for simple amplification and temperature-compensation circuits, where the number of pinouts is actually reduced from four to three for piezoresistive sensors.

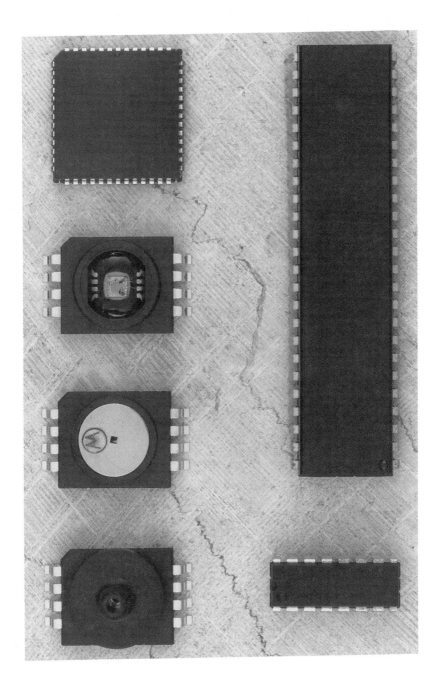

Figure 10.2 Pressure sensor and accelerometer packages compared to microcontroller packages (*courtesy of* Motorola, Inc.).

10.3 HYBRID PACKAGING

Hybrid packages such as ceramic multichip packages are routinely used during the research and development of semiconductors and MEMS devices. This allows researchers to probe selected portions of the die and verify expected functionality, especially if the entire die is not performing to predicted performance levels. Hybrid packages are also used for production sensors.

10.3.1 Ceramic Packaging and Ceramic Substrates

Ceramic packages, such as the CERDIP (or ceramic DIP), utilize a lead frame that is attached to the ceramic base through a glass layer. After die and wire bonding, a ceramic top is glass-sealed to the base. This same technique is used for other form-factor ceramic packages, including the ceramic flat pack. Ceramic packages are usually used for high-reliability applications and are much more expensive than other semiconductor packaging techniques. They are very useful in the development phase of a sensor since the silicon die does not have to be encapsulated. This allow various test points on the die to be easily probed and measured in packaged form [5].

Ceramic technology is used in hybrid assembly techniques for sensors. The ceramic substrate, usually an aluminum oxide, provides a firm mounting platform for the sensor die. Stress isolation can be obtained by utilizing a compliant silicone for the die attachment. The ceramic substrate allows laser trimming of thick-film resistors deposited on the ceramic surface to provide the calibration for a signal-conditioned sensor.

10.3.2 Multichip Modules

The extensive research that is being performed in the area of combined-technology microelectronics is also being evaluated and adapted for manufacturing combined sensor(s) and microelectronics. Figure 10.3 shows potential packaging techniques that could possibly be used for sensors [4]. These approaches include conventional chip-and-wire, flip-chip, and TAB. Chip-and-wire is the standard die-on-substrate packaging technique. Flip-chip will be discussed further in Section 10.4.3. TAB packaging eliminates wiring bonding from the die to a lead by directly attaching a lead to the top of the die. Any one of these methods is a potential candidate for MCM packaging.

For MCMs with one or more sensing elements and silicon ICs, silicon is one of the substrate materials that could be used. The approach of a "silicon circuit board" to achieve wafer-level hybrid integration may be the best interim solution for combined sensor and MCU. However, there are several varieties of MCMs, depending mainly on the substrate technology: MCM-L, MCM-C, MCM-D, and MCM-Si are common classifications [6].

An *MCM-L*, where the -L suffix stands for laminate, uses advanced printed circuit-board technology, copper conductors and plastic laminate-based dielectrics. This is es-

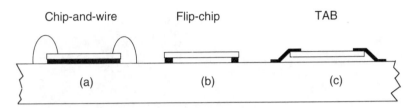

Figure 10.3 Bare die mounting techniques for multichip modules.

sentially a chip-on-board (COB) technology, as shown in Figure 10(a). An *MCM-C* uses thick-film screening on cofired ceramic substrates. An *MCM-D* has interconnections formed by the thin-film deposition of metals on dielectrics of polymers or inorganic compounds. An *MCM-Si* uses a silicon substrate with aluminum or copper interconnections and SiO_2 as the inorganic dielectric media.

The yields for MCMs can be lower than for assemblies made using packaged semiconductors. Recent known good die (KGD) efforts in the semiconductor industry are directed towards improving yields by improved testing at the wafer level. These same approaches will be required for MCM sensors.

10.3.3 Dual-Chip Packaging

Custom packages that accommodate a sensor die and its associated signal-conditioning circuit are common for high-volume applications. In these instances, the volume justifies the development and tooling costs. An example of a pressure sensor package developed for automotive manifold absolute pressure sensors is shown in Figure 10.4 [7]. Thin-film resistors on the control die are laser-trimmed to provide a fully signal-conditioned and calibrated pressure sensor. The control die has been used as a building block for other pressure applications and with different packaging requirements. The two-chip approach has proved to be both cost-effective and flexible.

10.3.4 Ball Grid Array Packaging

Ball grid array (BGA) packages are among the newer semiconductor packages that may also have potential for multichip smart sensors. A overmolded pad array carrier (OMPAC™) package is one of the new developments in semiconductor packaging, which also is known as solder bump array (SBA), pad array carrier (PAC), and land grid array (LGA). Each of these techniques relies on a blind solder joint for direct attachment of a leadless chip carrier to a printed circuit board [8]. As shown in Figure 10.5, the OMPAC has solder bumps attached to the copper foil pad on an epoxy glass substrate. The die in the plastic molded package is attached by gold wire bonds. Thermal vias run between solder

Figure 10.4 Dual-chip custom plastic package.

bumps to dissipate heat from the package. The solder bumps reflow during the surface-mount assembly process to attach the package to a printed circuit board.

10.4 PACKAGING FOR MONOLITHIC SENSORS

The highest volume semiconductor packaging techniques utilize molded plastic packages with form factors that include single in-line packages (SIP), dual in-line packages (DIP), quad flat packs, and a variety of surface-mount devices (SMD). The lead frame and molding techniques used in plastic packages provide the lowest cost semiconductor packages. Improvements that have occurred in recent years include improved capability to withstand autoclave testing for improved hermeticity and the use of improved, low-stress mold compounds for surface-mount applications. However, a sensor that must physically interface with a mechanical system, such as a pressure sensor, presents additional packaging challenges.

Cylindrical TO-5 and TO-8 metal-can semiconductor packages have been adapted

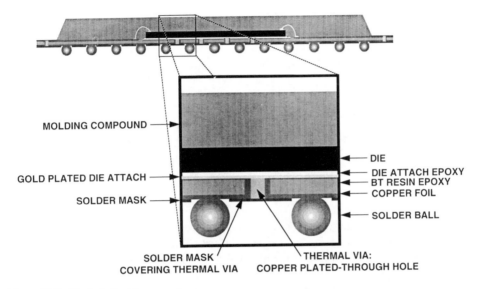

Figure 10.5 Plastic ball grid array package.

for use as sensor packages. For pressure sensors, the lid and header of the package usually have a hole to allow the pressure source to be applied to the die and provide a pressure reference to atmosphere. The leads extend through the base of the package through glass mounting seals. The use of gold plating material and hard die attachment to the glass substrate of the sensor provides one of the most media-compatible sensors. The metal lid is welded to the header and pressure sensor port attachment is accomplished through solder, brazed, or weld techniques [5]. These packages are also used for accelerometers, optical devices, and other sensors.

10.4.1 Plastic Packaging

A plastic chip carrier package that uses leadframe technology has been used for manu-facturing pressure sensors for over a decade. A single mold forms both the body and back of the chip carrier. The patented unibody package provides lower cost, fewer process steps, higher pressure range capability, and greater media compatibility when compared to earlier versions that were made of separate body and metal backplate [5]. The leadframe assembly technique allows easy handling of several devices at one time and the automation of assembly operations such as die-bond, wire-bond, and gel-filling operations. Automation allows tight process controls to be implemented and still provides high throughput.

One of the more difficult problems for sensors that interface to harsh environments is media compatibility. One approach to achieving a low-cost interface is illustrated in Figure 10.6.

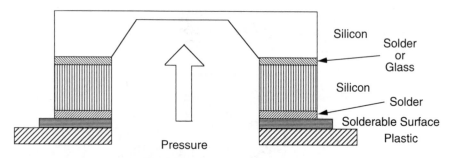

Figure 10.6 Increased media compatibility pressure sensor packaging.

10.4.2 Surface-Mount Packaging

The packaging shift to surface-mount technology indicated in Figure 10.1 is also affecting sensors. This can mean specifically designed surface-mounted sensor packages or, in more complex sensors, a printed circuit board (PCB) assembly that exclusively uses surface-mount technology. An example of the extent of surface-mount packaging in a sensor is demonstrated by the global positioning system sensor shown in Figure 10.7. This 2-in by 3 1/4-in by 1/2-in PCB has five surface-mounted ICs, including a 32-bit microcontroller unit [9].

The single-board radar sensor in Figure 10.8 shows several surface-mounted components [10]. As volume increases in these applications, increased integration and possibly higher density multichip modules will allow smaller packaging. High-volume manufacturing cost for a 0.5-in^2 version of the circuit in Figure 10.8 is estimated to be about $1.

10.4.3 Flip-Chip

Flip-chip packaging technology commonly used in integrated circuits is starting to receive attention in sensors [11]. Flip-chip technology will allow the electronics to be fabricated on one chip that can be attached to the MEMS chip through either a fluxless process or flux-assisted solder reflow process. This is a MCM-Si approach to combining different silicon processing technologies instead of increasing masking layers and die size to achieve a monolithic silicon solution.

An example of flip-chip technology in a standard IC process is shown in Figure 10.9(a) [12]. This technology has been used in automotive and computer applications for several years. The solder bump is formed over plated copper at the wafer level. Several steps precede the actual formation of the bump. A passivation nitride is used to mask areas other than the aluminum-copper metal contact on the silicon die. Sputtered TiW and copper are used to form a base metal for copper, lead, and tin, which are plated onto the base. Reflow in a gradually ramped oven causes a thermally induced amalgam of tin and lead to form into a ball due to surface tension. Spacing of the bumps evenly around the die avoids stress during thermal expansion.

Figure 10.7 Surface-mount packaging in Oncore™ global positioning system (*courtesy of* Motorola, Inc.).

Figure 10.9 also shows how flip-chip technology can be applied to a sensor. The sensor is represented by a diaphragm structure, but could be any micromachined structure. In one case, the sensor could be a flip-chip on a more complex IC such as an MCU. An alternate approach is to have the IC as a flip-chip on the micromachined structure. The requirements of the sensor must be considered when this type of packaging is developed.

10.4.4 Wafer-Level Packaging

Surface-micromachined structures sealed inside a two-layer bulk-micromachined sensor package were discussed in Chapter 2. This wafer-level "packaging" of micromachined sensors results in considerable cost savings over packaging methods that use metal can or ceramic packages. The process has demonstrated manufacturability and yields a well-protected hermetic sensor device such as an accelerometer. This technology has far-reaching implications and opens the door to new methods of low-cost hermetic packaging over the more expensive methods used today. Protection of the two-layer structure has been accomplished with a molded plastic package for current products. Future protective packaging may take a different form.

Epoxy protection for ICs has provided a low-cost packaging solution for consumer

Figure 10.8 Indoor radar sensor (*courtesy of* Lawrence Livermore National Laboratory).

electronics. The glob-top package shown in Figure 10.10(a) is simply an epoxy mound that protects the die surface and wire bonds for an IC mounted on a circuit board. For flip-chips, a protective layer around the die has been proposed that is shown in Figure 10.10(b) [13]. In this case, the protection is from silicon to silicon with sensitive layers sealed inside the epoxy. Lower cost packaging will be essential for many new smart-sensor applications. Adapting these approaches could provide the cost breakthrough.

10.5 RELIABILITY IMPLICATIONS

Sensors require tests similar to integrated circuits and unique qualification tests to verify that they will have acceptable performance of both the silicon and the packaging for their intended applications. For example, several tests have been developed for silicon pressure

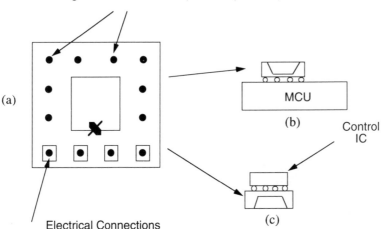

Mounting or Additional Circuit (i.e., Temperature) Connections

(a)

MCU

(b)

Control
IC

(c)

Electrical Connections

Figure 10.9 Flip-chip assembly (a) die with bumps, (b) bumped sensor on MCU, (c) bumped control die on sensor, and (d) bump structure itself.

sensors that are based on the need to detect potential failures due to the environment in which these devices will operate. These tests have been used to qualify sensors with on-chip amplification. Key tests include, but are not limited to [14]:

- Operational life, such as pulse, pressure-temperature cycling with bias;
- High humidity, high temperature with bias;
- High temperature with bias;
- High and low-temperature storage life;
- Temperature cycling;
- Mechanical shock;
- Variable frequency vibration;
- Solderability;
- Backside blow-off (for pressure sensors);
- Salt atmosphere.

These tests use accelerated life and mechanical integrity testing to determine the lifetime reliability statistics for silicon pressure sensors. Potential failure mechanisms are determined by the materials, processes, and process variability that can occur in the manufacturing of a particular sensor. For example, the pressure sensor of Figure 10.11 has 10 areas with 73 different items that can impact reliability [14]. Other sensors will have reliability test requirements that are commensurate with the application for which the sensor is being used and the type of packaging techniques that is employed.

The newest approach for establishing media compatibility for sensors uses in situ monitoring [15]. A typical three-step approach for verifying media compatibility is

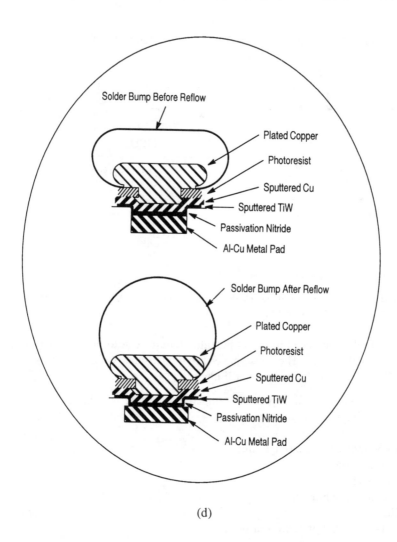

(d)

Figure 10.9 Continued.

(1) characterize, (2) expose to environment, and (3) retest for conformity to specification. This requires removing the sensor from the test environment and also changes the exposure of the sensor through the addition of oxygen prior to retest. New procedures have been developed that test the organic compounds used in the sensor's packaging, the silicon sensing element, and the integrated circuitry in the sensor, without removing the sensor from the test chamber. New sensor materials and modified packaging designs have been shown to survive in a variety of harsh media common to applications in automobiles and

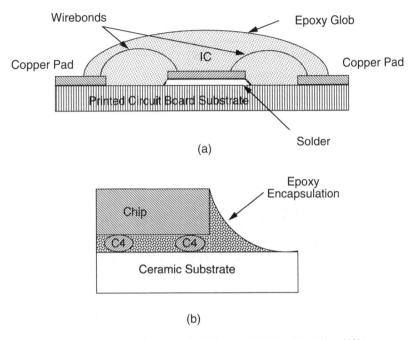

Figure 10.10 Epoxy protection for single die: (a) glob-top, and (b) flip-chip. (*After:* [13].)

trucks. Previously a comparable confidence level would have been possible only after achieving millions of unit hours in the actual application or experiencing a high rate of warranty returns.

10.5.1 Wafer-Level Sensor Reliability

One of the inherent advantages of semiconductor sensors is their capability of integrating the signal-conditioning, temperature compensation and essentially any aspect of semiconductor fabrication on the same silicon substrate as the sensor. This allows wafer-level reliability to be implemented. Test structures on the wafer or monitor wafers allow evaluation of critical process steps as they occur and provide rapid feedback to detect and correct process errors. The cost of testing at the wafer level is considerably less than testing at the completed assembly level. As shown in Table 10.2 [16], the cost of evaluating critical process steps early in the wafer fabrication process provides a considerable cost reduction over testing performed after the assembly process is completed. These data are a comparison of semiconductor process testing only. However, as sensor technology advances, packages such the as flip-chip and KGD approaches to packaging (common in the semi-

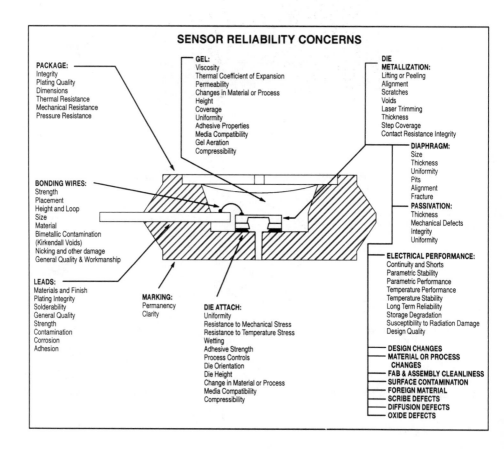

Figure 10.11 Reliability areas in a plastic-packaged pressure sensor.

conductor industry) will also be applied to sensors. Wafer-level reliability will have increased interest for both the sensor manufacturer and the user.

Normal test techniques require packaged units and considerably more time for evaluation. End of life techniques require even longer testing to ensure that a product life (e.g., 20 years) is achievable. For wafer-level testing, additional testing at the package level is required since packaging methods have a significant impact on the total reliability. However, wafer-level reliability testing is one of the ways that semiconductor sensors can continue to achieve lower product cost and continuous improvement in the future.

Table 10.2 Cost of Testing for Various Tests

| | | Test Technique | |
| | | Traditional | |
Test	Wafer-Level	Normal	End of Life
Gate-oxide evaluation	$200	$20,000	$32,000
Top-passivation evaluation	$175	$12,000	$16,000
New etch process	$800	$15,000	$13,000
Electromigration testing	$350	$ 7,000	$35,000
Hot-electron susceptibility	$750	$ 9,000	$ 8,500

After: [16].

10.6 SUMMARY

As smarter sensing technology is developed, microcontroller or application-specific integrated circuit (ASIC) capabilities will be included with the sensor. This will occur either as a monolithic structure or as multichip components, but in either case an increased number of pins will be added to the other sensor packaging requirements. However, all the justification for future cost-effectiveness of semiconductor-based sensors is predicated on these sensors following the learning curve similar to semiconductors. For sensors to approach the progress made in semiconductors, plastic packaging must be developed. Furthermore, packaging standards must be standardized for test equipment (handlers) and second sources. Sensor packaging draws heavily from both semiconductor and hybrid packages, but new techniques that are unique to smart sensors will most likely be developed to meet future requirements.

OMPAC and ONCORE are trademarks of Motorola, Inc.

REFERENCES

[1] Heitmann, R., "Ultra-Fine Pitch Technology: Assembly Challenges and Considerations," *Electronic Packaging & Production*, Dec. 1993, pp. 34–37.

[2] Van Zant, P., *Microchip Fabrication*, New York, NY: McGraw-Hill, Inc., 1990.

[3] Benson, A. F., "Count on Conformal Coatings," *Assembly Engineering*, June 1990, pp. 20–23.

[4] Frank, R., and J. Staller, "The Merging of Micromachining and Microelectronics," *Proc. of the Third International Forum on ASIC and Transducer Technology*, May 20–23, 1990, Banff, Alberta, Canada, pp. 53–60.

[5] Ristic, Lj., *Sensor Technology and Devices*, Norwood, MA: Artech House, Inc., 1994.

[6] Blood, W. R., and J. S. Carey, "Practical Manufacturing of MCMs," *Surface Mount Technology*, Nov. 1992, pp. 16–22.

[7] Czarnocki, W. S., and J. P. Schuster, "Robust, Modular, Integrated Pressure Sensor," *Sensor 95 International Conference on Sensors, Transducers, and Systems*, Nuremberg, Germany, May 1995.

[8] Chin, S., "Ball Grid Array Package Challenges Quad Flatpack," *Electronic Products*, April 1993, pp. 19–20.

[9] Oncore Ad, *Automotive Industries*, Oct. 1994, p. 71.

[10] Babyak, R. J., "Electronics," *Appliance Manufacturer*, May 1994, pp. 95–99.

[11] Markus, K. W., V. Dhuler, and A. Cowen, "Smart MEMS: Flip Chip Integration of MEMS and Electronics," *Proceedings of Sensors Expo*, Cleveland, OH, Sept. 20–22, 1994, pp. 559–564.

[12] *Linear/Interface ICs Device Data Book*, Vol. II, DL128/D, Rev. 4, Motorola Semiconducor Products Sector, Phoenix, AZ, 1994.

[13] Hill, G., J. Clementi, and J. Palomaki, "Epoxy Encapsulation Improves Flip Chip Bonding," *Electronic Packaging and Production*, Aug. 1993, pp. 46–49.

[14] Maudie, T., and B. Tucker, "Reliability Issues for Silicon Pressure Sensors," *Proc. of Sensors Expo '91*, Chicago, IL, Oct. 1–3, 1991, pp. 101-1–101-8.

[15] Frank, R., and T. Maudie, "Surviving the Automotive Environment," *Electronic Engineering Times 1995 Systems Design Guide*, Special Issue, 1995, pp. 62–63.

[16] Reedholm, J., and T. Turner, "Wafer-Level Reliability," *Microelectronics Manufacturing Technology*, April 1991, pp. 28–32.

Chapter 11

Mechatronics and Sensing Systems

"People are fascinated by what they can't quite understand."
—Michael Palin of *Monty Python's Flying Circus*

11.1 INTRODUCTION

To achieve the highest level of smart sensing, essentially the entire system will become a sensing system or an extension of sensing capability. The term mechatronics is frequently used to describe subsystem design for a combination of electromechanical components. Computers or microcontrollers are an integral part of many mechatronic systems. Mechatronics has been defined as the synergistic combination of precision mechanical engineering, electronic control, and the systems approach for designing products and manufacturing processes [1]. Mechatronics represents a higher level of sophistication and complexity, requiring designing at the interface of the electromechanical system. In some ways, mechatronics is similar to analog-to-digital interface—only more complex. Although engineers from different disciplines share their results on a particular program, each engineer works with a less than perfect understanding of how his/her design decision affects decisions of his/her counterpart. Mechatronics and other approaches, such as concurrent engineering, are attempting to understand design tradeoffs earlier in the design process in order to reduce design cycle time and ensure that the final product meets the design criteria.

11.1.1 Integration and Mechatronics

The need for increased control capability in electronic systems is driving increased levels of integration in semiconductor electronics, including sensors. A fully integrated silicon system is frequently cited as a goal for the ultimate system. Silicon integration is certainly progressing towards this capability, especially for selected portions of systems. However, the cost and performance capabilities of higher levels of integration must compete against the components that they would displace. Also, design flexibility and use in several applications must be considered. This includes diverse requirements such as satisfying both domestic and international requirements, or simultaneously meeting cost-sensitive low-end specifications and also providing feature-sensitive high-end products. Furthermore, safety and reliability concerns, such as the potential failure modes from a system failure mode and effect analysis (FMEA) can lead to more distributed intelligence when single-point failures cannot be tolerated. In some cases, the system constraints make it desirable to have sensing and power control elements remote, or at least separated, from the microcontroller that is controlling the system. This allows improved power dissipation, fault detection, and protection from system voltage extremes [2].

System partitioning and the determination of how much integration should occur within each component are decisions that should be made by an interactive design team composed of both system and silicon designers. Integration for the output section of the control system is frequently called a smart-power IC. These integrated solutions have been used in volume production by several manufacturers since the mid 1980s. Understanding their contribution to the control system and how they take advantage of embedded sensing in both normal operation and fault modes is essential to achieve the systems approach and implement mechatronics methodology. This chapter will examine smart-power ICs, more complex sensing in arrays, and systems aspects of mechatronics.

11.2 SMART-POWER ICS

Smart-power ICs and power ICs (PICs) typically combine bipolar and MOS circuitry with power MOSFET technology to provide direct interface between the MCU and system loads, such as solenoids, lamps, and motors. Smart-power ICs are defined by the Joint Electron Devices Engineering Council (JEDEC) as

> [H]ybrid or monolithic devices that are capable of being conduction-cooled, perform signal conditioning, and include a power control function such as fault management and/or diagnostics. The scope [of this definition] shall apply to devices with a power dissipation of 2W or more, capable of operating at a case temperature of 100°C and with a continuous current of 1A or more.

Smart-power devices also provide increased functionality as well as sophisticated diagnostics and protection circuitry. Sensing of current levels and junction temperature is a key aspect of the design for control during normal operation and for detecting several fault modes.

The power in smart-power ICs is typically a power MOSFET (or DMOS, for diffused MOS device). The size of this portion of the silicon area is determined by the process technology, voltage rating, and desired on-resistance. Process technology determines the on-resistance per unit area and has provided improvements based on reducing the cell size and improving cell geometry. The voltage rating inversely affects efficiency—a higher voltage rating means a higher on-resistance for a given area of the power device and lower efficiency. For a given process and voltage rating, the number of cells is increased to meet on-resistance specification. Since increased silicon area directly impacts the cost, the highest on-resistance to allow safe power dissipation in the system is normally specified.

The smart-power approach to system design means that a number of circuit elements can be consolidated into one single device. These devices would previously have been discrete components, or a combination of standard or custom IC and discrete output devices. As illustrated in Figure 11.1, input, frequently from a microcontroller, initiates the startup or turnon procedure. The control can have circuitry for a soft start function or charge pump circuitry for increasing the gate drive voltage. A high degree of functionality is achieved by smart-power ICs, due to the interaction between the control circuit, power device, protection circuitry, and the load. Multiple power devices on one single chip are the most cost-effective uses of this technology. This provides space saving, component reduction, total system cost reduction, improved performance, and increased reliability from the reduced number of interconnections.

The choice of process technology for smart-power ICs has historically depended upon the type of control elements that were integrated. Some circuit elements, like operational amplifiers (op amps), comparators, and regulators are best implemented using a bipolar IC process. MOS circuitry handles logic, active filters (time delays), and current mirrors better than bipolar circuitry. Some circuits, such as analog to digital converters (ADCs) or power amplifiers, can be implemented equally well in either technology. Today, a process that has both MOS and bipolar for the control circuitry does not have to sacrifice performance

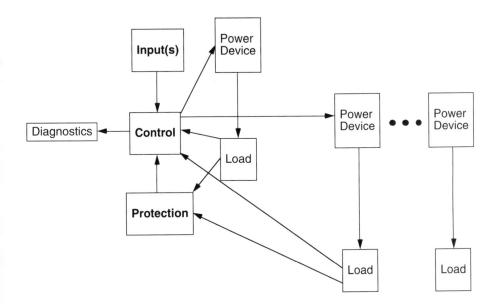

Figure 11.1 Smart-power IC block diagram.

or features and, if it is combined with the appropriate output devices, can handle the power control functions for a number of system loads. However, just as a smart sensor needs to fully utilize the capability of a more expensive process, a smart-power IC must also provide advantages beyond those that separate control IC and discrete power devices can provide. One of the key means to achieving increased functionality is to integrate sensing into the smart-power IC.

11.3 EMBEDDED SENSING

Sensing in smart-power ICs detects fault conditions and threshold conditions, and this allows implementation of control strategies. Overtemperature, overcurrent, and overvoltage conditions can be sensed, and the smart-power IC can be designed to react or simply provide a signal to the MCU for system-safe operation. The different sensing techniques and applications of the sensed signal will be explored in this section.

11.3.1 Temperature Sensing

Temperature sensors can be easily produced in semiconductor devices by using either the temperature characteristics of the base-emitter voltage of a transistor or a diffused resistor. In addition, circuitry is easily integrated to produce a monolithic temperature sensor with an output that can be easily interfaced to a microcontroller. By using an embedded temperature sensor, additional features can be added to ICs. (Note: a (temperature) sensor becomes an embedded item in a semiconductor product when it has a secondary instead of primary function.)

Sensing for fault conditions, such as a short-circuit, is an integral part of many smart-power ICs. The ability to obtain temperature sensors in a mixed bipolar-CMOS-DMOS semiconductor process provides protection and diagnostics as part of the features of smart-power ICs. The primary function of the PIC is to provide a microcontroller-to-load interface for solenoids, lamps, and motors. In multiple output devices, sensing the junction temperature of each device allows the status of each device to be input to the microcontroller and, if necessary, the MCU can shut down a particular unit that has a fault condition [3].

In ICs, the temperature limit can be sensed by establishing a bandgap reference. Two diodes' junctions are biased with different current densities, but the ratio is essentially constant over the operating temperature range (-40°C to 150°C). The following equation shows how the differential voltage (ΔVbe) is related to the current (I) and emitter area (A) of the respective transistors.

$$Vbe1 - Vbe2 = \Delta Vbe = [k_B T/q] \cdot \ln\left[\frac{I1/A1}{I2/A2}\right] \tag{11.1}$$

where k_B = Boltzman's constant (1.38×10^{-23} joules/K), T = temperature in K, and q = charge of electron (1.6×10^{-19} C).

At a targeted high temperature, the negative temperature coefficient of *V*be equals the positive temperature coefficient voltage developed across a sense resistor, and a thermal limit signal is generated. Other circuitry regulates the voltage and turns off the output power device.

A smart-power IC can have multiple power drivers integrated on a single monolithic piece of silicon. Each of these drivers can have a temperature sensor integrated to determine the proper operating status and shut off only a specific driver if a fault occurs. Figure 11.2 shows an eight-output driver that independently shuts down the output of a particular driver if its temperature is excessive (between 155°C and 185°C).

The octal serial switch (OSS) adds thermal sensing through overtemperature detection circuitry to the protection features. Faults can be detected for each output device and individual shutdown can be implemented. In a multiple-output PIC it is highly desirable to shut down only the device that is experiencing a fault condition and not all of the devices that are integrated on the PIC. With outputs in various physical locations on the chip, it is difficult to predict the thermal gradients that could occur in a fault situation. Local temperature sensing at each output, instead of a single global temperature sensor, is required.

As shown in Figure 11.3, the eight outputs of the device with individual temperature sensors can be independently shut down when the thermal limit of 170°C is exceeded. All of the outputs were shorted to 16V supply at a room-temperature ambient. A total current

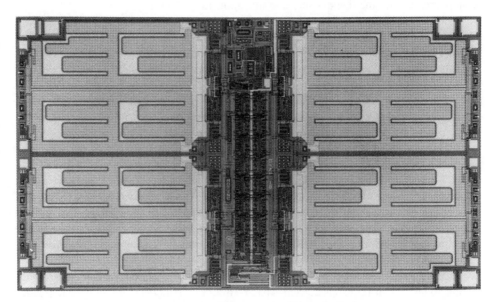

Figure 11.2 Photomicrograph of eight-output PIC (*courtesy of* Motorola, Inc.).

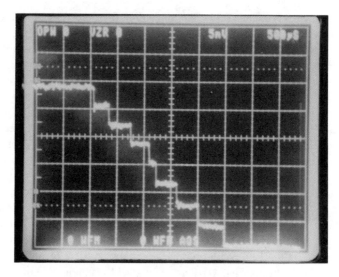

Figure 11.3 Independent thermal shutdown of an eight-output smart-power IC.

of almost 30A initially flowed through the device. Note that each device turns off independently. The hottest device turns off first. Variations can result from differences in current level and thermal efficiencies. As each device turns off, the total power dissipation in the chip decreases and the devices that are still on heat up slower.

This hard short could have been diagnosed by short-circuit (current limit) sensing. However, a soft short, which by definition is below the current limit but exceeds the power-dissipating capability of the chip, can be an extremely difficult condition to detect. Soft shorts require overtemperature sensing to protect the IC from destructive temperature levels.

The overtemperature condition sensed by the power IC may mean that the device turns itself off to prevent failure in one case; in another situation, a fault signal provides a warning to the MCU but no action is taken, depending on the fault circuit design. The remaining portion of the system is allowed to function normally. With the fault conditions supplied to the MCU, an orderly system shutdown can be implemented.

Temperature sensing has also been integrated into a standard power MOSFET process. The power MOSFET process was chosen instead of a power IC process because a large die size was required to meet the power dissipation requirements of a very cost sensitive automotive application. A minor modification (the addition of a single masking layer) to a production power MOSFET process allows both sensing and other protection features to be integrated.

The thermal sensing is provided by integrated polysilicon diodes. By monitoring the output voltage when a constant current is passed through the diode(s), an accurate indication of the maximum die temperature can be obtained. A number of diodes are actually provided in the design. A single diode has a temperature coefficient of 1.90 mV/°C. Two or more can be placed in series if a larger output is desired. For greater accuracy, the diodes can be trimmed during the wafer probe process by blowing fusible links made from polysilicon.

The response time of the temperature-sensing diodes is less than 100 μs, and this has allowed the power device to withstand a direct short across an automobile battery. The sensor's output was applied to external circuitry that provided shutdown prior to device failure. The sensing capability also allows the output device to provide an indication (with additional external circuitry) if the heatsinking is not proper when the unit is installed in a module or if a change occurs in the application that would ultimately cause a failure.

11.3.2 Current Sensing in Power ICs

Current sensing provides protection and control in power ICs. A device that demonstrates this approach is shown in Figure 11.4 [4]. The circuit (Figure 11.4(a)) includes a voltage regulator, oscillator, and charge pump to provide a highside switch with an N-channel power MOSFET. This custom device has multiple current-sensing levels for control and fault detection (Figure 11.4(b)). The current-sensing technique uses a sample of the current flowing through the total power MOSFET cells (the ratio of the number of sample cells to control cells) as an indication of the total current. A resistor in series with the sense cells, which handle considerably lower current, provides a voltage proportional to the current in the power device. This SENSEFET™ technique avoids adding a series resistance to the total on-resistance and the subsequent power consumption, which could be excessive for an integrated circuit. The isolated sensing cells for four different current sensing levels are inside the four "lassoed" areas in Figure 11.4(b). The highest current level could trigger an overcurrent fault and the lower levels can be used for system-specific control points.

11.3.3 Diagnostics

The eight-output driver with serial communications shown in Figure 11.2 represents a high level of complexity for diagnostics. Control and sensing of the faults of individual output drivers could require over 24 connections for eight-output drivers. By using a serial peripheral interface (SPI) protocol described in Chapter 6, 8-bit serial control of the output and independent diagnostics are possible with only three connections. The SPI operates at a frequency of up to 2 MHz and utilizes the low-voltage high-speed CMOS capability in the SMARTMOS™ (Motorola's power IC) process. With both the input and the output tied directly to the MCU, the unit acts as a closed-loop subsystem between the load and the MCU for each load. By using a daisy chain technique, several additional devices can be added to the control capability. A total of 32 outputs are controlled with only four connections (the IRQ, interrupt request not, is optional) to the MCU.

A high level of smart-power IC device complexity is demonstrated in a parallel-serial control. This device has six outputs that can be selected independently on six dedicated

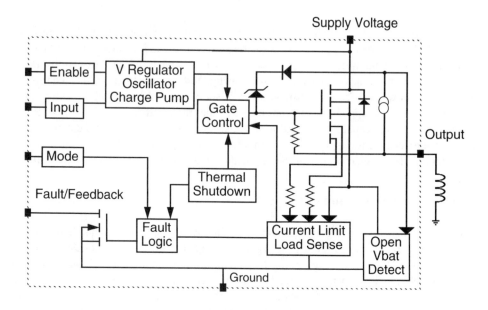

(a)

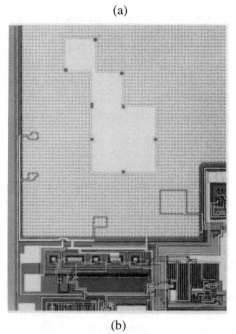

(b)

Figure 11.4 Current sensing integrated in smart-power IC: (a) block diagram, and (b) die photomicrograph (*courtesy of* Motorola, Inc.).

parallel input pins by an input command (voltage above the threshold voltage) from an MCU. Inputs 0, 1, 4, and 5 can also be simultaneously selected. This allows the two output devices to be connected in parallel for reducing the on-resistance. Parallel control provides the fastest activation of the loads, which allows real-time control using pulse-width modulation (PWM) techniques. The drawback is the highest number of connections must be made and the highest number of traces must be routed on the printed circuit board. Serial control of the six outputs is also possible using a SPI interface, as described earlier. Mixed control of the outputs can be achieved with various combinations of parallel and serial control of the outputs.

Fault diagnostics and control are similar to those in the octal serial switch. However, an advanced SPI interface provides serial diagnostic information to the MCU, which includes immediate parity checking (n_{th} word) to confirm that the smart-power IC received the word sent by the MCU. In the initial SPI implementation (the OSS), fault reporting is performed in the $n - 1$ word. The present command to the OSS is compared to the outputs from the previous word to determine the status. The time of the active filter period, which can be several clock cycles, is required before the next word can determine if a fault exists. The advanced SPI uses an exclusive OR to provide fault information when the current word is written.

Automotive applications have required diagnostics beyond those typically found in smart-power ICs. For example, the California Air Resource Board's latest on-board diagnostics (OBDII) legislation requires that the actual movement of the pintle in transmission solenoids be monitored when the output device is activated. This is done to indicate that the transmission is in the proper gear and that the torque converter has full lock-up. Simply sensing the current going through the output device is insufficient to determine that the pintle has moved as a result of the output device being turned on. The differential sensing circuit in Figure 11.5(a) is used to detect the change that occurs when the solenoid's armature is attracted into the winding. The negative slope, shown in the current waveform in Figure 11.5(b) is an indication of the physical movement. Either lowside or highside sensing can be performed with lowpass or highpass filters and a comparator providing the signal conditioning. The event counter can be adjusted from 15 to 64 to establish the confidence level of detecting movement. The on-resistance of the power driver can be used as a sense resistor to convert the current into a voltage level. A peak detector can also be used to detect the peak in Figure 11.5a. The results can be reported on a serial SPI or with parallel fault reporting. The OBDII diagnostic adds one more aspect to the fault sensing/detection capability of smart-power ICs. A summary of the faults that can be detected is shown in Table 11.1 [2].

Fault reporting techniques and corresponding MCU commands to the smart-power IC can be accomplished by several methods. For new devices, the most appropriate communication technique is dictated by the application requirements. The communication techniques that are available as building blocks are summarized in Table 11.2 [2].

For distributed control systems, both smart-sensing and smart-power nodes are required. Smart-power ICs combined with an MCU or MCUs with integrated power devices

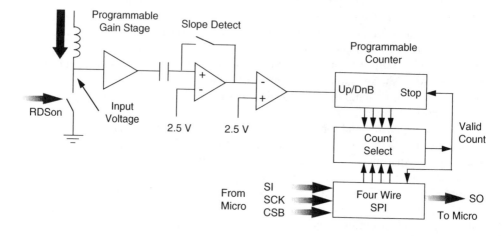

(a)

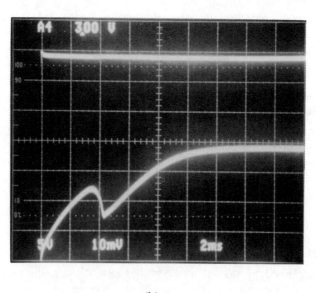

(b)

Figure 11.5 Detecting solenoid activation (a) circuit, and (b) waveform.

Table 11.1 Fault Diagnostics in Automotive System

Fault	Technique
Open load (on-state)	Logic comparison
Open load (on- or off-state)	Logic comparison
Shorted load	Overcurrent sensing
Overvoltage (load dump)	Overvoltage sensing
Overtemperature (single output)	Temperature sensing
Independent overtemperature (multiple output)	Temperature sensing
Solenoid movement	Bipolar + logic

Table 11. 2 Summary of Communication Techniques for Power ICs

Parameter	Communication	Capability
Input	Parallel	Within slew rate (typ $\leq$ 1A/μs)
	Serial	2-MHz SPI
	Parallel/serial	Selectable/mixed
Diagnostic	Flag (single)	Open/shorted load
	Flag (single)	Open (on- or off-state)/shorted load
	Serial	2-MHz SPI (all faults)
	Flag (multiple with interrupt)	Reports all faults

implement the load control. This means that the power or output side of the system must also be able to communicate using the system protocol. The protocols discussed in Chapter 6 must take into account the power portion of the system. For a power node to be added to an existing system, it must have the appropriate protocol to operate on the system.

11.4 SENSING ARRAYS

More than one sensor is frequently required in order to provide sufficient information for a control system. R&D efforts are progressing in several areas to integrate the sensors ultimately on the same silicon wafer with signal conditioning, computational capabilities, and possibly even actuation. In the short term, a single-package solution is proving to be an improvement over prior units. The sensing arrays can include a number of sensors for different measurands (e.g., pressure, flow, temperature, and vibration). Multiple sensors of a given type can be used to increase the range, provide redundancy, or capture information at different spatial points. Also, multiple sensors of a given type, with minor modifications, can be used to measure different species in chemical sensing applications.

11.4.1 Multiple Sensing Devices

Improving the performance of sensors can be accomplished by simultaneously fabricating several sensors on the same silicon substrate or mounting several sensors in a hybrid package. One example of the monolithic approach is the surface-micromachined pressure sensor developed by the Fraunhofer Institute of Microelectronic Circuits and Systems [5]. Surface micromachining was combined with CMOS processing to achieve a monolithic smart sensor. A single absolute capacitive surface-micromachined pressure sensor with a 100-µm diameter membrane provided approximately a 0.017-pF capacitance without applied pressure. Using up to 81 individual pressure sensors, which were switched in parallel, capacitance values between 1 to 2 pF were achieved. The sensor output varied approximately 0.2 pF over a 1 to 6 bar pressure range.

A multichannel probe designed for measuring single-unit activity in neural structures is shown in Figure 11.6 [6]. Eight active recording sites are selected from 32 sites on the probe shank using a static input channel selector. A large number of recording sites is desirable for sampling the total activity within even a restricted tissue volume. In this design, on-chip CMOS circuitry amplifies and multiplexes the recorded signals and electronically positions the recording sites with respect to the active neurons. The neuron signals typically have frequency components from about 100 Hz to 6 kHz, with an amplitude from 0 to 500 µV. The on-chip functions of the probe include channel selection, amplification, signal bandlimiting, multiplexing, clocking, power-on reset, and self-testing.

Multiple sensing sites are common in chemical measurements. A multielement smart gas analyzer has been developed that represents one of the most sophisticated sensors with respect to both operation and data processing [7]. The multielement sensor has a number of thin-film detectors to overcome the drawbacks of an earlier version and to improve selectivity and specificity in analyzing gas mixtures.

A four-element gas analyzer is achieved by a three-chip hybrid configuration. Figure 11.7 shows a block diagram of the hybrid gas analyzer. Two dual-window gas detectors and a control interface chip are in the hybrid package. The gas detector process is

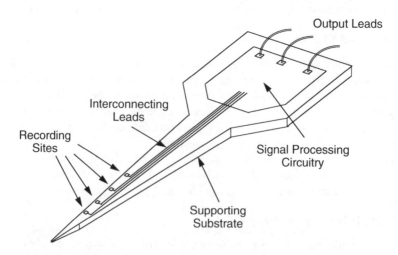

Figure 11.6 Multichannel microprobe sensor. (*After:* [7].)

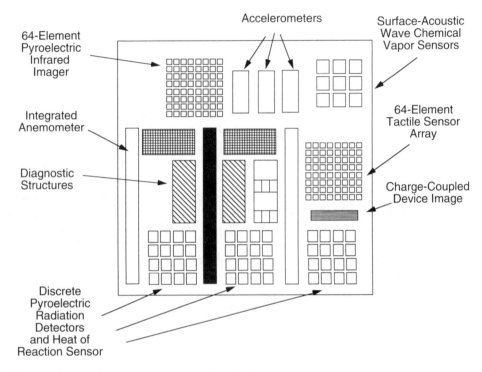

Figure 11.7 Multiple-type integrated sensor. (*After:* [12].)

compatible with CMOS processing, so the combination could eventually be integrated into a single monolithic chip.

The control chip is capable of controlling four gas-detecting elements independently. All elements can be monitored and programmed simultaneously. A temperature of 1,000°C can be controlled for each chip. The chip communicates with a local node MCU or a remote processor through an eight-pin standard interface. The front-end standard is capable of operating up to 32 sensors, 32 actuators, 32 self-test commands, and 32 special-function commands for different applications.

Another array approach for gas sensors has a 3 by 3 matrix of identical sensors on a single 4,466 by 6,755-μm chip [8]. The nine sensors are interdigitated gate electrode field-effect transistors or IGEFETs. A single version of this design has selectively and reversibly detected 2 parts per billion of nitrogen dioxide and diisopropyl methylphosphonate—two pollutants that can adversely affect the environment. By using a Fourier transform signal-processing technique and implementing a pattern recognition algorithm, it is possible for this type of sensor to operate as an electronic nose for detecting and identifying the constituents of a multicomponent gas mixture.

Photodiode arrays are becoming cost-effective for manufacturing inspection, quality control, process monitoring, and other industrial applications. A silicon charge-coupled device (CCD) has a spectral range from 200 to 1,100 nm. Devices can be manufactured with detector elements as small as several microns in width and with up to 2,048 elements. After amplification and digitizing for signal processing, the readout can routinely reach more than 1 MHz. A complete spectral measurement can be made in only a few milliseconds. Optical design, electronic design, and mechanical packaging must be coordinated in a mechatronics methodology to allow these devices to be used in laboratory as well as manufacturing environment [9].

An infrared focal plane array that operates much like the human retina has been designed and fabricated [10]. The array, called a Neuromorphic sensor, has logarithmic sensitivity to avoid saturation common in conventional thermal imagers. It also performs pixel-based sensor fusion and real-time local contrast enhancement. The sensor contains a backside illuminated 64 by 64 array of 100-μm indium antimony photovoltaic diodes. A CMOS IC performs the readout function as well as two-dimensional averaging across the chip. The Neuromorphic sensor mimics the neural network process by interconnectivity between each pixel's four nearest neighbors. A digital computer performing the same functions at the same processing rate as the Neuromorphic sensor would require approximately 100W . The Neuromorphic device consumes only 50 mW or nearly 1/2,000 of the power. This is a very dramatic demonstration of the difference that smart sensors will contribute to future systems.

11.4.2 Multiple Types of Sensors

A variety of sensors have been fabricated on a single chip to detect damage and performance degradation caused during semiconductor assembly and packaging operations [11]. The sensors were designed using CMOS technology common to many IC manufacturing facilities. Sensors on the chip include an ion detector, a moisture sensor, an electrostatic discharge detector, a strain gauge for measuring shear forces, an edge damage detector, and a corrosion detector as well as heaters for corrosion, acceleration and thermal modeling.

Figure 11.7 shows another example of an integrated multiple-sensor chip that utilizes the pyroelectric and piezoelectric effects in zinc oxide thin films [12]. Sensors on the chip include a gas flow sensor, an infrared-sensing array, a chemical-reaction sensor, cantilever beam accelerometers, surface acoustic wave (SAW) vapor sensors, a 64-element tactile sensor array, and an infrared charge-coupled device imager. The 8- by 9-mm^2 chip also has MOS devices for signal conditioning, array accessing, and output buffering. Backside micromachining of the silicon is performed as the last step in the fabrication process. Low processing cost per function was demonstrated by the multielement sensor.

11.4.3 An Integrated Sensing System

An example of a fully integrated sensing system using multiple sensing elements is a monolithic hearing aid. The hearing aid requires only two functional elements to be a system: the sensor and the amplification. A micromachined piezoelectric microphone with on-chip CMOS signal conditioning has been designed and fabricated that has potential for hearing-aid applications [13].

The sensor consists of eight piezoelectric (zinc oxide) electrode pairs. Bulk micromachining to define the microphone diaphragm is performed prior to CMOS processing. The design takes into account a 9-hour 1,150°C anneal and stresses imposed by subsequent CMOS processing steps. After CMOS circuitry is completed, the 0.5-μm thick zinc oxide is magnetron-sputtered onto the diaphragm. The eight pairs of upper and lower electrodes are connected in two series groups of four electrode-pairs. The microphone has a measured sensitivity of 0.92 mV/Pa and a resonant frequency of 18.3 kHz. Preamplifier noise of only 13 μV allows the microphone to achieve an A-weighted noise level of 57-dB sound pressure level (SPL), which is roughly the level of conversational speech.

11.5 OTHER SYSTEM ASPECTS

The smart-sensor system requires additional components to provide a self-contained unit. Current developments would allow batteries and fluorescent displays to be integrated with the sensor. Any item mentioned in Chapters 8 and 9 may also be part of this integration—if the packaging problems discussed in Chapter 10 are solved. These components, combined with smart sensors, digital logic from an MCU or DSP, and smart-power ICs, will enable a number of new mechatronics approaches. One system aspect that must be addressed by one of these areas is the catastrophic effect that electrostatic discharge (ESD), system voltage transients, and EMI can have on electronic components. These will be covered in the last section in this chapter.

11.5.1 Batteries

Portable applications have generated considerable system design activity to obtain extended performance and improved life from the batteries that power these systems. By including semiconductor components in the battery, a smart battery is created. A smart battery is a battery or battery pack with specialized hardware that provides information regarding the state of charge and calculates predicted run time [14]. Information regarding the type of battery chemistry and the battery-pack voltage, capacity, and physical packaging is conveyed by a protocol to power-related devices in the system. The system can provide control to a smart battery or batteries, a smart-battery charger, and various regulators and switches. Other battery-charging efforts are directed at eliminating the memory effect in nickel-cadmium (NiCd) batteries, reducing the charging cycle time, and avoiding overcharging, which reduces battery life. Portable data acquisition systems and remote sensing devices will benefit when smart batteries are incorporated into their design.

Reduced battery size through thin-film manufacturing technology will also impact sensors. A thin-film solid-state Li/TiS_2 microbattery has been reported that has properties suitable for long-life rechargeable applications [15]. The microbatteries range in size from 8 to 12 μm in thickness and have a capacity between 35 and 100 μA-hrs/cm^2. The open-circuit voltage is approximately 2.5V. Batteries have been cycled over 10,000 cycles at 100 μA/cm^2. Microbatteries routinely withstood 1,000 cycles between 1.4 and 2.8V at current densities as high as 300 μA/cm^2.

Any reasonably smooth surface is a potential substrate for the thin-film battery. A chromium, TiS$_2$, and solid electrolyte layer are sputtered onto the substrate. LiI, Li, and a protective coating are vapor-deposited over the sputtered layers. The construction technique will allow the microbattery to be incorporated with many semiconductor devices, including microsensors, during their manufacture.

11.5.2 Field Emission Displays

Efficient low-power displays are an essential part of many systems, including portable digital assistants, virtual-reality-driven robots, and automotive global-positioning systems. These displays communicate information from the machine to the operator. Currently, active-matrix liquid crystal displays (LCDs) are the dominant technology, but flat panel displays (FPDs)and an emerging technology called field emission displays (FEDs) are potential replacements in several areas. FEDS produce light using colored phosphors. They do not need complicated, power-consuming backlights and filters, and almost all light is visible to the user. In addition, no power is consumed by pixels in the off state. An FED consists of an array of microtips that are micromachined into a substrate. Metal, silicon, and diamond are being investigated for the microtips [16]. Successful implementation of any one of these approaches will provide additional possibilities for a sensor-to-operator interface.

11.5.3 System Voltage Transients, ESD, and EMI

Surviving voltage transients, electrostatic discharge (ESD), and electromagnetic interference (EMI) are common application requirements, especially in the automotive environment. The ability to withstand a load dump is the prime factor in defining the voltage capability of semiconductor devices used in automotive applications. The load-dump transient is a high-voltage (> 100V in an unsuppressed system) high-energy transient generated by disconnecting the battery from the alternator when the engine rpm is high and the alternator is generating a high-level output. Techniques of handling load dump include (1) squelching the energy from an excessively high voltage input (> 60V) by turning on an active clamp until the energy decays (few hundreds of milliseconds), and (2) designing the output devices to withstand high voltage without turning on. The latter approach is done with an increase in the resistance of each output device and therefore causes more power dissipation under normal operating conditions. This technique must be used for high-voltage (> 100V) pulses to avoid excessive power dissipation during the load dump. A third alternative uses the circuitry in the smart-power device to turn the output on for a period of time (about 80 ms) when an overvoltage occurs. The input is time filtered before sampling to determine if the gate drain clamp should be turned on again [2].

Semiconductor sensor designs and applications have a system requirement that is normally not a problem in presemiconductor mechanical sensors, namely, ESD. Automotive and other applications generate high levels of ESD during manufacturing or in service. As a result, automotive manufacturers specify some of the most stringent ESD testing to qualify components. Automotive module manufacturers are indicating that future specifications will require any IC pin that is accessible outside the module to withstand a 10-kV electrostatic discharge with 250 pF and a 1,500-ohm load. This means that peak currents approaching 7A and energy levels in the 12.5 to 14-millijoules (mj) range must be tolerated. The output connections of sensors, power, and control circuitry can be exposed to electrostatic discharge during assembly or operation [2].

Sensors must be designed to meet levels of EMI commonly found in the application or required by industry specifications. EMI is a disturbance or malfunction of equipment or systems caused by the operation of other equipment or systems or by the forces of nature. Distributed control systems can be susceptible to high-frequency switching that is common in computers, switching power supplies, and power control. Figure 11.8 shows an industrial

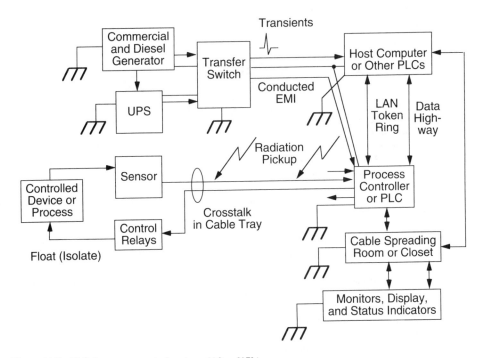

Figure 11.8 EMI in process control system. (*After*: [17].)

process control with ground loops and other EMI paths [17]. Sensors can be easily affected, depending on the wiring. Isolating the wiring, avoiding ground loops, and in some cases uses shielded wiring, may be necessary to avoid problems.

In automotive applications, some manufacturers require the sensor to withstand electric fields of magnitudes up to 200 V/m in the engine compartment and up to 50 V/m in the passenger compartment [18]. Performance levels and measurement methods are defined in a number of Society of Automotive Engineers (SAE) documents including SAE J551, SAE J1816, SAE J1113, SAE J1407, and SAE J1338. Classifications of failure-mode severity are provided in SAE J1812.

11.6 SUMMARY

The control of the output portion of the system made possible by smart-power ICs has been discussed. This technology combined with smart sensors and computing technology is part of mechatronics—a holistic approach to the system. New developments in batteries and display technology will help to redefine the "sensor." Advances in all of these areas will contribute to new generations of smarter sensing and smart-sensing systems that are interactive sensing networks—especially if potential system problems are taken into account early in the design process.

SENSEFET and SMARTMOS are trademarks of Motorola, Inc.

REFERENCES

[1] Comerford, R., "Mecha...What?" *IEEE Spectrum*, Vol. 31, No. 8, Aug. 1994, pp. 46–49.

[2] Wollschlager, R., K. M. Wellnitz, and R. Frank, "Diagnostics and Communications in Vehicle Control Systems Using Smart Power ICs," SAE SP-1013, *Sensors and Actuators 1994*, Warrendale, PA, pp. 21–27.

[3] Frank, R., "Embedded Temperature Sensors for Protection and Control," *Sensors*, May 1995, pp. 38–45.

[4] Himelick, J. M., J. R. Shreve, and G. A. West, "Smart Power for the 1990s in General Motors Automobiles," *ISATA 22nd International Symposium on Automotive Technology & Automation*, Florence, Italy, May 14–19, 1989.

[5] Kandler, M., J. Eichholz, Y. Manoli, and W. Mokwa, "Smart CMOS Pressure Sensor," *Proc. of the 22nd International Symposium on Automotive Technology & Automation* #90185, Florence, Italy, May 14–18, 1990, pp. 445–449.

[6] Ji, J., and K. D. Wise, "An Implantable CMOS Analog Signal Processor for Multiplexed Microelectrode Recording Arrays," *IEEE Solid-State Sensor and Actuator Workshop*," Hilton Head, SC, June 4–7, 1990, pp. 107–110.

[7] Najafi, N., "A Multi-Element Gas Analyzer Utilized in a Smart Sensing System," *Proc. of Sensors Expo West*, San Jose, CA, March 2–4, 1993, pp. 97–109.

[8] Kolesar, E. S., "Sensitive and Selective Toxic Gas Detection Achieved with a Metal-Doped Phthalocyanine Semiconductor and the Interdigitated Gate Electrode Field-Effect Transistor (IGEFET)," *Proc. of Sensors Expo*, Cleveland, OH, Sept. 20–22, 1994, pp. 11–30.

[9] Pelnar, T., "Photodiode Array Spectra Analysis and Measurements for Industry," *Sensors Expo West*, San Jose, CA, March 2–4, 1993, pp. 337–339.

[10] Massie, M., "Neuromorphic Sensor Mimics the Human Eye," *EDN Products Edition*, Jan. 17, 1994, p. 34.

[11] Yates, W., "Test Chips Detect Semiconductor Problems," *Electronic Products*, Sept. 1991, pp. 17–18.

[12] Polla, D. L., and R. S. Muller, "Integrated Multisensor Chip," *IEEE Electron Device Newsletter*, Vol. EDL-7, No. 4, April 1986, pp. 254–256.

[13] Ried, R. P., E. S. Kim, D. M. Hong, and R. S. Muller, "Piezoelectric Microphone with On-Chip CMOS Circuits," *IEEE Journal of Microelectromechanical Systems*, Vol. 2, No. 3, Sept. 1993, pp. 111–120.

[14] Maliniak, D., "Smart Batteries Spelled Out by Joint Enabling Specification," *Electronic Design*, June 13, 1994, p. 50.

[15] Jones, S. D., J. R. Akridge, "Microfabricated Solid-State Secondary Batteries for Microsensors," *Proc. of Sensors Expo*, Cleveland, OH, Sept. 20–22, 1994, pp. 215–223.

[16] Derbyshire, K., "Beyond AMLCDs: Field Emission Displays?" *Solid State Technology*, Nov. 1994, pp. 55–65.

[17] White, D., "Electromagnetic Compatibility: What is it? Why is it Needed?" *Instrument & Control Systems*, Jan. 1995, pp. 65–74.

[18] *SAE Handbook*, Vol. 2, "Parts and Components," Warrendale, PA, 1995 or latest edition.

Chapter 12

The Next Phase of Sensing Systems

"Once you have eliminated the impossible, whatever is left *is possible.*"
—Ancestor of Mr. Spock, Chief Science Officer, Starship Enterprise

12.1 INTRODUCTION

As Chapter 11 pointed out, the combination of several techniques can be used to develop the next generation of smart-sensing systems. This has been possible for many years and is indeed the basis of many instruments such as the Fabry-Perot interferometer, blood testing systems, and the camera on a chip. What makes the future possibilities exciting is the scale (size) of the sensors made possible by the combination of micromachining and microelectronics. The size of semiconductor sensors is directly proportional to the cost of the sensor, and the cost is inversely proportional to the volume of applications. When cost barriers are broken, laboratory curiosities become a part of everyday life. Manifold absolute-pressure sensors in cars, disposable blood-pressure sensors for medical applications, and accelerometers for air bag systems have demonstrated the extent that a cost-effective sensing technology can displace a previous technology. This is especially true when on-chip integration is possible.

In the areas of computing, communications, and entertainment the convergence of these technologies is on the verge of creating entirely new products and markets. Similarly, computing, communications, and other technologies will be combined with several or all of the topics that have been discussed in this book to create new control products. These new products may require the combinations of several enabling technologies that are in the process of achieving manufacturing status. Aggressive semiconductor industry roadmaps, heavy investment in R&D for sensors, and the large number of participants in sensor technology and manufacturing promise a variety of competitive products with the focus on the smart aspects of these sensors.

12.2 FUTURE SEMICONDUCTOR CAPABILITIES

Since micromachined sensors are based on semiconductor technology, a semiconductor industry roadmap could provide insight into future sensing technology. The Semiconductor

Industry Association (SIA) recently published technology predictions to the year 2010. As shown in Table 12.1 [1], by 2010 the minimum feature size for the highest performance microprocessor circuits will be only 0.07 μm—1/5 the size of a 1995 feature. The cost will be reduced by an order of magnitude more than the feature size dropping from 1 to 0.02 millicents per transistor. The number of logic transistors will increase by over 22 times, from 4 million to 90 million. Unfortunately, there are no direct predictions about micromachining technology or sensors. However, the part that makes any sensor smart—the computational engine—is increasing at a phenomenal rate. The supercomputer that perform today's laboratory calculations will be a personal computer with similar performance at some point in the future.

Sensing technology will play an important role in developing the next generation of semiconductor technology. Sandia National Laboratory has developed a silicon chip with up to 250 microsensors to monitor mechanical, chemical, and thermal environment of integrated circuits [2]. This chip can be used during prototyping, manufacturing, or anytime during the life of the chip to monitor critical parameters that could affect performance. The assembly test chip with its onboard polyimide and Al_2O_3 moisture sensors, piezoelectric strain gauges, electrostatic discharge and corrosion detectors, mobile ion detectors, and thermocouples promises to play an important role in developing semiconductor technology.

The scale of future semiconductors will require measuring and monitoring techniques that are well beyond today's capability—especially if they will be used on a day-to-day basis to monitor production process and measure quality control. The use of contamination sensors for measuring and monitoring particulates and moisture are among the recommendations from the SIA roadmap [3]. Candidates for sensors include low-cost gas-analysis sensors such as residual gas analysis, optical emission spectroscopy, and intracavity laser spectroscopy. The integration of these sensors into online equipment was recommended. Metal and total-oxidizable-carbon contamination measurements in liquids were also considered essential. The report frequently cites the need to monitor wafer contamination during processing as a key to achieving manufacturable product as the critical dimensions

Table 12.1 Projected Semiconductor Technology Capability

Characteristic	1995	1998	2001	2004	2007	2010
Minimum feature size (in μm)	0.35	0.25	0.18	0.13	0.1	0.07
Logic transistors/cm^2 (packed)	4M	7M	13M	25M	50M	90M
Cost/transistor @ volume (in millicents)	1	0.5	0.2	0.1	0.05	0.02
Maximum no. of wiring Levels (logic)	4–5	5	5–6	6	6–7	7–8
Electrical defect density(d/m^2)	240	160	140	120	100	25
Minimum mask count	18	20	20	22	22	24
Chip size (mm^2) ASIC	450	660	750	900	1100	1400
Power supply voltage (desktop)	3.3	2.5	1.8	1.5	1.2	0.9
No. of chip I/Os	900	1,350	2,000	2,600	3,600	4,800
Performance (in MHz) (chip-board)	150	200	250	300	375	475

shrink. Undoubtedly, sensors that allow semiconductor manufacturers to achieve the predicted capability will be pursued and implemented as they are available. The Sematech (a consortium of semiconductor manufacturers) project to develop interoperable sensor and actuator standards confirms the industry's desire to communicate the information from these sensors in an intelligent or smart manner using a common protocol [4].

12.3 FUTURE SYSTEM REQUIREMENTS

The unprecedented capability that semiconductor technology will provide raises other questions. How much technology will we really need? Who is going to apply this technology to smart sensors? The answers to these questions and several others that could be asked come from examples that are present today.

The extensive use of manifold absolute-pressure sensors in vehicle emission control systems resulted from two factors: (1) government legislation forced a change in the way automotive engines were controlled, and (2) semiconductor-based sensors proved to be more cost-effective and more reliable than the mechanical versions that were originally used to define the control system. Legislation and customer demand drove the development of technology.

In invasive blood-pressure measurements, disposable semiconductor-based sensors initially proved to be lower cost than resterilizing and periodically recalibrating expensive mechanical units. As the replacement of mechanical units was preceding at an "as predicted" rate, the AIDS virus appeared and accelerated the conversion process. Cost-effectiveness and consumer demand drove this second sensor application.

More recently, government legislation mandated passive restraint systems for the driver in passenger vehicles. The manufacturers' choices were an automatic seat belt or air bag system. Customers quickly accepted the air bag system after numerous reports of users walking away from accidents that would have previously been classified as fatal crashes. As a result, an application that was opposed by vehicle manufacturers in the 1970s quickly became a selling feature in the 1990s. Relative to the cost of the vehicle, an air bag system was perceived by buyers as a reason to buy one vehicle over another, and in some cases justification for buying a new vehicle because their old vehicle just was not as safe. The mechanical crash sensors initially used in these systems did not have a self-test feature and required calibration for each body style. Semiconductor-based accelerometers have allowed the number of crash sensors to be reduced from as many as five to only one or two, depending on the system's redundancy requirements. Furthermore, the system can be tuned to meet the requirements of a particular body style after the sensor is in the module by programming the EEPROM in the microcontroller. Also, several techniques have been developed by sensor manufacturers that allow the accelerometer to be activated for self-testing each time the vehicle is started. The ability to perform the self-test and verify that this critical input to the system is capable of performing its designed function any time during the life of the vehicle is comforting to both the driver and the front-seat passenger in vehicles equipped with these systems. Once again, legislation and customer demand have driven semiconductor sensor development and acceptance.

Now, the obvious question is where are the next opportunities for semiconductor-based sensors to meet a legislated requirement and/or customer demand? The answer is revealed in part by the technology that is already being developed. Providing additional safety or security and reducing energy consumption and emissions are behind many of the potential applications in today's sensor R&D labs. Legislated requirements always provide a guaranteed market and focus development efforts on a specific application. Legislation has been enacted for fugitive emissions; CO detection in homes, trailers, and recreational vehicles; and reduced energy consumption in industrial, building, and home applications. Applications addressing an aging population of baby boomers with discretionary funds to spend on entertainment and the avoidance of aging effects (hearing loss) are potential candidates for volume sensor usage. The question that must be answered now is can these semiconductor sensors provide cost-effective value-added functionality to the system? This can occur in an application where the sensor replaces a previous technology and allows more users to take advantage of an existing system based on cost reduction. Another possibility is a new system that can be designed based on the capability provided by semiconductor sensors combined with other new technologies. The examples in the next section demonstrate research that could be the next enabling technology for sensors.

12.4 NOT-SO-FUTURE SYSTEMS

Many of the future sensors that are produced will not surprise industry watchers because they were demonstrated by researchers at early stages of the technology development. The list at the end of this paragraph covers some of the more intriguing possibilities. In Chapter 9, Table 9.1 listed MEMS technology that could also produce manufacturable products at some point in the future. Examples in this section include a laboratory tool that can help researchers as well as technologies that may be used directly by a large number of consumers. The following is a list of sensors in development:

- Nose on a chip;
- Gas analysis system;
- Camera on a chip;
- Atomic force microscope;
- Monolithic hearing aid;
- Mass spectrometer;
- Vehicle dynamics sensor;
- Viscometer on a chip.

12.4.1 Fabry-Perot Interferometer

Fabry-Perot interferometers (FPI) are used in the laboratory to optically measure lengths or changes in length with a high degree of accuracy. A micromachined version consists of two silicon wafers with deposited highly reflective dielectric mirrors as shown in Figure 12.1 [5]. This is already a second-generation version of an earlier development. The FPI incorporates a number of critical micromachined elements and semiconductor processes, including the controlled gap, movable silicon mesa corrugated support, silicon fusion bonding, capping wafer, control electrodes, and optical coating. Miniature FPIs have applications in telephony systems to link large numbers of customers to a central exchange. In addition, a small-gap FPI may have applications in sensors to measure small displacements and provide the feedback element in a servo loop. Two visible-ultraviolet range interferometers in series are expected to give narrow-bandwidth selectivity for use as a spectrometer [6].

12.4.2 HVAC Sensor Chip

Improved energy efficiency and energy management are major driving forces for new products. The energy consumption is directly achieved by using improved efficiency power devices and control techniques such as pulse width modulation (PWM). The control techniques allow variable-frequency drive motors to be implemented in building automation. Buildings in the U.S. account for 35 to 40% of the nation's energy consumption.

All of the MEMS devices for a heating, ventilation, and air conditioning (HVAC) system, may eventually be integrated into a single device such as the one shown in Figure 12.2 [7]. In addition to the sensors, signal conditioning to amplify the signal, calibrate offset

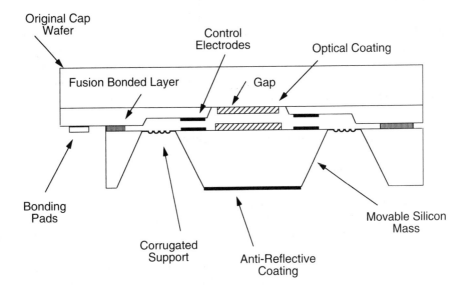

Figure 12.1 Fabry-Perot interferometer. (*After:* [5].)

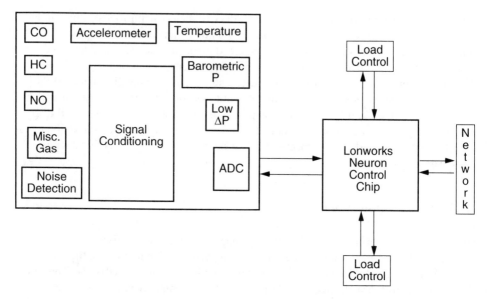

Figure 12.2 A highly integrated sensor solution for automated building control.

and full scale output, and compensate for temperature effects could be included on this device. Since it most likely will use CMOS technology for the semiconductor process, an onboard analog to digital converter (ADC) could also be integrated. Other system capability could include fuzzy logic and neural network interpretation of the input signals. This is especially true if an array of chemical sensors is used to indicate a wide variety of chemical species and overcome many of the problems of available chemical sensing products. Currently, sensors are not this highly integrated and are specified on an "as required" basis. This is due to the additional cost that they add to the initial system installation—and their inability to function without requiring more maintenance than other parts of the system.

12.4.3 Speech Recognition and Micro-Microphones

Combinations of technologies that have been discussed will provide future products. One possibility is the combination of speech recognition algorithms, complex computing capability, and micromachined microphones. Neural networks are being investigated for intelligent sensors in speaker-independent speech recognition systems [8]. This research has resulted in an acoustic preprocessing chip that offloads the computational requirements from the host computer, allowing a slower, more cost-effective computer to perform speech recognition. This portion of the system combined with a micromachined microphone such as the one developed by the University of California, Berkeley, could provide portable speech recognition and possibly language translation capability [9]. Of course, both portions of these systems still require considerable development effort to be manufacturable as separate elements. However, when they can be manufactured, the combination of these devices and the resulting benefits will become obvious to many designers.

12.4.4 Microgyroscope

Small cost-effective gyroscopes are being developed to measure pitch, roll, and yaw in vehicle control systems. Gyroscopes are part of the sensing inputs for intelligent transportation system (ITS) navigation controls and for combined antilock braking and traction control systems. One proposal uses silicon micromachining to produce a three-layer structure for sensing angular rate [10]. As shown in Figure 12.3, the surface-micromachined structure has the first and third layers fixed and the second layer is free to rotate around its axis. The center is held in place by four spring support arms attached to four mounting posts. The sensor measures rotation around the X and Y axes, and it senses acceleration in the direction of the Z axis. The comb drive provides oscillation about the Z axis. Differential capacitive sensing detects angular rotation as a displacement of the vibrating disk. The sensing of the three functions is performed by CMOS circuitry that can be integrated on the same chip as the gyroscope. This combination of micromachining and microelectronics will reduce equipment that previously cost hundreds to thousands of dollars to a cost level that allows the gyroscope to be embedded in a vehicle control system for use by consumers.

12.4.5 MCU with Integrated Pressure Sensor

A monolithic microcontroller with a micromachined pressure sensor has been proposed [11]. The key elements of this design are shown in Figure 12.4. The MCU/sensor combination is based on a widely available MCU architecture, with an integral bulk-micromachined sensor. Fabricating the chip is only one aspect of taking it from concept to reality. The packaging challenges discussed in Chapter 10 must also be addressed. At present, a separate MCU and sensor packaged at a module level provides a more cost-effective solution. However, a specific application may require the space savings or performance advantages that the monolithic solution can provide. At that point, the Level V sensor shown in Figure 1.7, the fully integrated sensing system, will become a reality.

12.5 SOFTWARE, SENSING, AND THE SYSTEM

Computer-aided design (CAD) and simulation capability is essential in both the electrical and the mechanical portions of the sensor. It is also becoming a critical element in any system design. Design tools that implement mechatronic methodology are being standardized that will produce a common level of understanding. Based on the increased

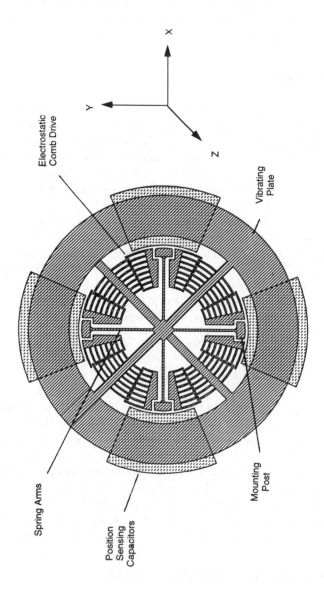

Figure 12.3 Three-layer gyroscope.

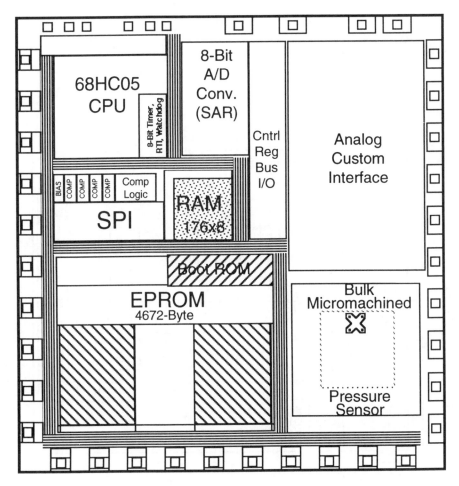

Figure 12.4 Microcontroller with an integrated pressure sensor.

capability that semiconductor hardware is providing, a more structured methodology will be the only way that designers will be able to cope with the semiconductor capability that will be available in the future. The number of transistors, the increased I/O possibilities, and the added difficulty inherent in high-density geometries will pose challenges to the most experienced designer. Add to this brief list only one more item, such as lower supply voltages, and it is easy to see that the system designer must start to deal with system problems at a higher, and earlier stage, in the design.

An example of the ideal *top-down methodology* is provided by automobile control systems and is shown in Figure 12.5 [12]. The full extent of interdependency and all of

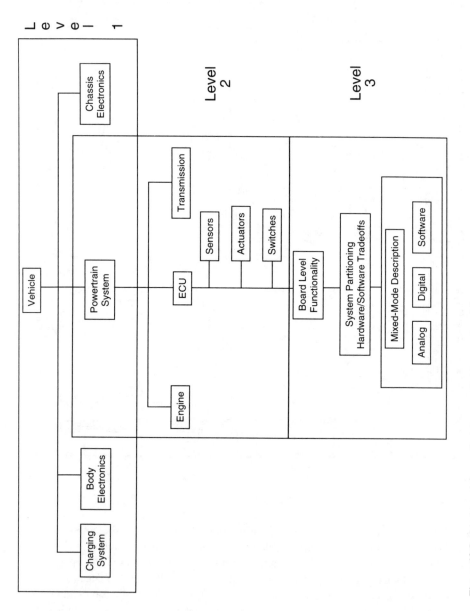

Figure 12.5 Ideal top-down methodology applied to an automobile.

the vehicle's control systems are not indicated in the figure. In this approach, the vehicle's electrical, electronic, and mechanical systems are modeled and simulated for interdependency at the beginning of the design process. Three levels can be defined that allow the semiconductor technologies (including the sensors) to be partitioned and the software to be developed. At the highest level (Level 1), the interrelation between the vehicle and its electronic subsystems must be modeled. Inside a particular subsystem (Level 2), such as the powertrain electronics, the engine, transmission, and electronics that control these mechanical systems must be modeled. At this level, the various components (sensors, actuators, and switches providing vehicle, environment, and driver input) are taken into account. Finally, at Level 3 the board-level functionality of the system, including silicon partitioning and the hardware and software aspects, must be addressed. It is at this level that the amount of intelligence that goes into the sensor and the amount that remains in the microcontroller must be determined.

To further complicate the designer's task, the control system consists of mechanical, digital, and analog elements. The recent availability of tools that address these areas from a top down methodology is crucial. Hardware descriptive languages (HDL) that address the digital and analog requirements have been developed by several suppliers in either Verilog A or VHDL-A. Tools that support mechatronics design are available to simulate the electrical and mechanical aspects of the system.

The system designer must evolve into both a generalist and yet a specialist in several areas to provide input to sensor and semiconductor suppliers and implement the actual system-level design. This is a challenge to a different way of thinking and to a different skill set than many of today's designers practice. System tradeoffs must be determined prior to releasing requests for quotes (RFQ) to suppliers, and, in fact, the supplier (or suppliers) must be involved in making design tradeoff decisions to achieve proper system partitioning and avoid unnecessary costs. Management must set priorities in a different way to achieve the initiatives that are required to implement this change in methodology. This is probably the most difficult challenge. It involves the commitment of additional design resources and tools at the beginning of the design cycle—when the level of commitment is lowest and the ability to delay or cancel the project is normally at its highest. Advanced design tools must be purchased that allow the modeling and system simulation to be performed prior to the availability of electronics hardware [12].

Software for sensor designers is already viewed as essential. Modeling for the structures and materials in the micromachining process has been the focus of universities and an integral part of the newest products provided by the sensor industry [13, 14]. These activities combined with mechatronics and mixed-signal modeling, which is gaining acceptance in the electronics design community, will provide the critical tools for smart sensing.

Supporting industry standards that provide volume incentive to suppliers, and therefore encourage cost competition, are a challenge to manufacturers, sensor suppliers, and electronic system suppliers. The combination of knowledge-based characterization of the system and standards provide the foundation to achieve world class manufacturing capability for sensors. It is the manufacturing capability that will take sensors from the researchers' laboratories into the lives of consumers.

12.6 ALTERNATE VIEWS OF SMART SENSING

The smart sensing definition in Chapter 1 is currently a proposal by an IEEE committee that will most likely be approved with only minor modifications, if any. However, as the technology has been evolving, other definitions have been offered that show a different view of smart-sensing technology and offer insight to what researchers think is in the future for smart sensors. Two examples will be presented.

One view of smart sensing proposes that a "smart sensor is defined as one that is capable of (i) providing a digital output; (ii) communicating through bidirectional digital bus; (iii) being accessed through a specific address; and (iv) executing commands and logical functions" [15]. The smart sensor also has desirable functions such as compensation of secondary parameters (e.g., temperature). failure prevention and detection, self-testing, autocalibration, and various computationally intensive operations. The emphasis on computing attributes in this definition necessitates a microcontroller as a minimum requirement for a smart sensor.

Another proposed definition states "a smart sensor consists of a transducer combined with other signal conditioning and other circuitry to produce output information signals that have been logically operated on to increase the value of the information to the system" [16]. The aspects of added signal conditioning and increased information value are noteworthy in this definition. The author lists sensor characteristics that take several sensors, from the simplest to smartest version. Figure 12.6 shows these characteristics.

It is doubtful that either of these two authors would be satisfied with the definition proposed by the IEEE. However, the key to smart sensing will be in the ability to satisfy users, not those making definitions. Sensor manufacturers will apply the term smart sensor to a product even if it does not meet an industry definition. But customers selecting those sensors that directly impact the performance and value of their product will determine the truly smart sensors. It is hoped that this book has in some manner provided guidance to manufacturers and users of smart sensors to understand their potential in control systems.

12.7 SUMMARY

Smart sensors will mean more than the IEEE definition presented in Chapter 1. However, technology is not smart if it is not cost-effective, and the much too frequently used "smart" prefix is used to describe anything that is an improvement over a previous version of a technology, product, or service. Examples of smart things include SmartTrading[SM] by OLDE, Smart Scrub[TM] from Dow, Smart Ones[TM] from Weight Watchers®, the Smart-Rate[SM] from the Discover[TM] Card, Smart Solutions[SM] from the United States Postal Service,

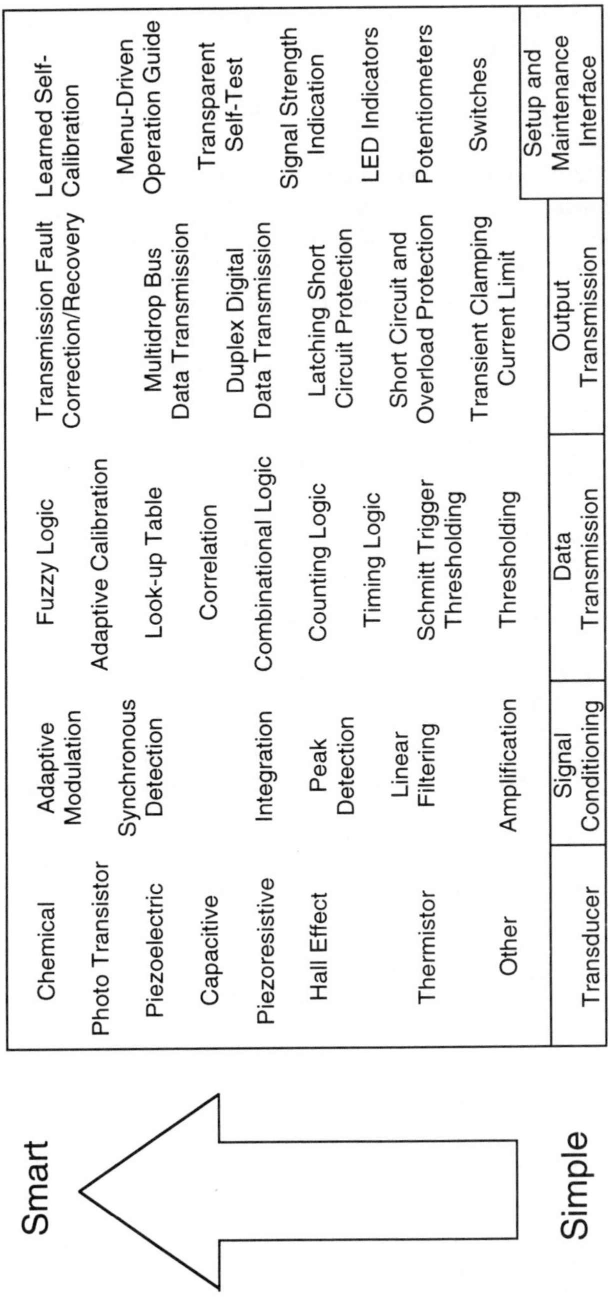

Figure 12.6 Smart sensor relative IQ.

and even Smart, Very Smart™ products from Magnavox. Smart is the late 20th century term that is used instead of ''new and improved version.''

Recent and projected advances in semiconductor technology and design tools will enable tomorrow's engineers to design, simulate, and verify something as complex as an entire vehicle system and probably to ''virtually'' prototype the entire vehicle. To effectively utilize these tools, the design methodology and even the designer's skill set must be modified to implement a top down methodology and the systems approach. Standards will play a critical role in reducing costs for sensor manufacturers that can be passed on to systems manufacturers and, ultimately, to consumers. Those aspects of sensing that were once thought to be impossible are enabling future smart sensors that will be limited only by the imagination.

All trademarks and service marks are the property of their respective owners.

REFERENCES

[1] *The National Technology Roadmap for Semiconductors*, Semiconductor Industry Association, San Jose, CA, 1994.

[2] Studt, T., ''Smart Sensors Widen View on Measuring Data,'' *R & D Magazine*, March, 1994, pp. 18–20.

[3] Haystead, J., ''SIA Roadmap Highlights Contamination Control,'' *Cleanrooms*, Vol. 9, No. 2, Feb. 1995, pp. 1, 10.

[4] Stock, A. D., and D. R. Judd, ''Interoperability Standard for Smart Sensors and Actuators Used in Semiconductor Manufacturing Equipment,'' *Proc. of Sensors Expo*, Philadelphia, Pa, Oct. 26–28, 1993, pp. 203–222.

[5] Jerman, J. H., D. J. Clift, and S. R. Mallinson, ''A Miniature Fabry-Perot Interferometer with a Corrugated Silicon Diaphragm,'' *Technical Digest of IEEE Solid-State Sensor and Actuator Workshop*, Hilton Head Island, SC, June 4–7, 1990, pp. 140–144.

[6] Raley, N. F. et al., ''A Fabry-Perot Microinterferometer for Visible Wavelengths,'' *Technical Digest of IEEE Solid-State Sensor and Actuator Workshop*, Hilton Head Island, SC, June 22–25, 1992, pp. 170–173.

[7] Frank, R., and D. Walters, ''MEMS Applications in Energy Management,'' *Sensors Expo*, Boston, MA, May 16–18, 1995, pp. 11–18.

[8] Donham, C. et al., ''A Neural Network Based Intelligent Acoustic Sensor,'' *Proc. of Sensors Expo*, Philadelphia, PA, Oct. 26–28, 1993, pp. 121–127.

[9] Ried, R. P. et al., ''Piezolectric Microphone with On-Chip CMOS Circuits,'' *Journal of Micromechanical Systems*, Vol. 2, No. 3, Sept. 1993, pp. 111–120.

[10] Dunn, W., ''Automotive Applications of Low G Accelerometers and Angular Rate Sensors,'' *Proc. of Sensors Expo*, Cleveland, OH, Sept. 20–22, 1994, pp. 195–201.

[11] Benson, M., R. Frank, J. Jandu, and M. Shaw, ''Advanced Semiconductor Technologies for Smart Sensors,'' *Proc. of Sensors Expo*, Philadelphia, PA, Oct. 26–28, 1993, pp. 133–143.

[12] Momin, S., and R. Frank, ''Automotive Electronics: The Challenges and Benefits of Top Down Design for Vehicle Manufacturers,'' *Symposium on Indian Automotive Technology*, Pune, India, Dec. 7–10, 1994, pp. 121–126.

[13] Zhang, Y., S. B. Crary, and K. D. Wise, ''Pressure Sensor Design and Simulation Using the CAEMEMS-D Module,'' *Technical Digest of IEEE Solid-State Sensor and Actuator Workshop*, Hilton Head Island, SC, June 4–7, 1990, pp. 32–35.

[14] Harris, R., M. F. Maseeh, and S. D. Senturia, ''Automatic Generation of a 3-D Model of a Microfabricated

Structure," *Technical Digest of IEEE Solid-State Sensor and Actuator Workshop*, Hilton Head Island, SC, June 4–7, 1990, pp. 36–41.

[15] Najafi, K., "Smart Sensor," *Journal of Micromechanics and Microengineering*, June 1991, pp. 86–102.

[16] Juds, S. M., "Towards a Definition of Smart Sensors," *Sensors*, July 1991, pp. 2–3.

Smart Sensor Acronym Decoder and Glossary

accuracy (also see error) a comparison of the actual output signal of a device to the true value of the input. The various errors (such as linearity, hysteresis, repeatability and temperature shift) attributing to the accuracy of a device are usually expressed as a percent of full-scale output (FSO).

actuator the part of an open-loop or closed-loop control system that connects the electronic control unit with the process.

adaptive (control) system of advanced process control that is capable of automatically adjusting itself to meet a desired output despite shifting control objectives and process conditions or unmodeled uncertainties in process dynamics.

AHDL analog hardware description language.

algorithm a set of well-defined rules or processes for solving a problem in a finite number of steps.

aliasing distortion due to sampling a continuous signal at too low a rate.

aliasing noise distortion component created when frequencies present in a sampled signal are greater than one-half the sample rate.

ALU arithmetic logic unit; the unit of a computing system containing circuits that perform arithmetic and logical operations.

analog output an electrical output from a sensor that changes proportionately with any change in input.

anneal heat process used to remove stress, especially in surface micromachining.

anisotropic etching that is dependent on crystallographic orientation.

ANSI American National Standards Institute.

---

antialiasing filter normally a lowpass filter that band-limits an input signal before sampling to prevent aliasing noise.

APK amplitude-phase keying; a modulation technique for RF signal transmission.

arbitration process of gaining access to bus.

architecture also, system architecture-the hardware or software design, usually standardized, for a circuit or system.

ASIC application-specific integrated circuit; an IC designed for a custom requirement, frequently a gate array or programmable logic device.

ASK amplitude shift keying; a modulation technique for RF signal transmission.

ASME American Society of Mechanical Engineers.

associative memory a neural network architecture used in pattern recognition applications in which the network is used to associate data patterns with specific classes or categories it has already learned.

ASTM American Society for Testing and Materials.

ATPG automation test program/pattern generator.

attenuation decrease in magnitude of communication signal.

autozero control circuitry that periodically re-establishes the zero to avoid drift errors.

bandgap reference forward-biased emitter junction characteristics of transistor used to provide an output voltage with zero temperature coefficient.

bandpass filter filter designed to transmit a band of frequencies while rejecting all others.

baseband frequency band occupied by information-bearing signals before combining with a carrier in the modulation process.

baud unit of signaling speed equal to the number of discrete signal conditions or events per second. Refers to the physical symbols/second used within a transmission channel.

BAW bulk acoustic wave.

BIST built-in self-test; design technique that allows a chip to be tested for a guaranteed level of functionality.

bit rate speed at which data bits are transmitted over a communication path, usually expressed in bits per second (bps). A 9,600-bps terminal is a 2,400-baud system with 4 bit/baud.

bridge forwards packets of information between channels in multiple media network.

bulk micromachining a process for making microstructures in which a masked (silicon) wafer is etched in orientation-dependent etching solutions.

bus connection between internal or external circuit components.

calibration a process of modifying sensor output to improve output accuracy.

C4 controlled collapse chip connection; solder bumps on IC for circuit connection.

CCITT Consultive Committee of the International Telephone and Telegraph.

CDMA code-division multiple access; spread-spectrum method of allowing multiple users to share the RF spectrum by assigning each active user an individual code.

CDPD cellular digital packet data; wide area data network that takes advantage of existing AMPS (U.S.) cellular network by transmitting data packets on unused voice channels. Data is transmitted using RS (63, 47) at an effective rate of 14 Kbps.

chip a die (unpackaged semiconductor device) cut from a silicon wafer, incorporating semiconductor circuit elements such as a sensor, actuator, resistor, diode, transistor, and/or capacitor.

CISC complex instruction set computer; standard computing approach as compared to RISC architecture.

closed loop control system that utilizes a sensing device for measuring a process variable and making control decisions based on that feedback.

CMOS complementary metal oxide semiconductor.

COB chip on board packaging technique for semiconductor die in multichip modules.

codec code-decode; translates audio to digital signals and digital back to audio signals, usually with an A/D and D/A converter.

collision state of a data bus in which two or more transmitters are turned on simultaneously to conflicting states.

combinational technologies integrated mixed-signal (analog and digital) technology.

common-mode rejection (ratio) CMR or CMRR; the ratio of the common-mode input voltage to output voltage commonly expressed in dB (i.e., the extent to which a differential amplifier rejects an output when the same signal is applied to both inputs).

companding the process of reducing the dynamic range of voice or music signal (compressing), typically at the transmitter, and then restoring (expanding) at the receiver.

compensation added circuitry or materials designed to counteract known source of error.

contention ability to gain access to the bus on a predetermined priority.

convolution mathematical process that describes the operation of filters.

CPU central processing unit; the portion of the computer that includes the circuits that control the interpretation and execution of instructions.

CSMA/CD carrier zense multiple access with collision detection; the access technique used in the Ethernet protocol.

cyclic redundancy check (CRC) an error-detecting code in which the code is defined to be the remainder resulting from dividing the bits to be checked in the frame by a predetermined binary number.

data compression technique that provides for the transmission of fewer data bits than originally required without information loss. The receiving location expands the received data bits into the original bit sequence.

data logging method of recording a process variable over a period of time.

DCS distributed control system.

decimation process by which a sampled and digitized data stream is modified in order to extract relevant information.

Delta-P ΔP; change in pressure or pressure differential.

deterministic the property of a signal that allows its future behavior to be precisely predicted.

DFT design for testability; methodology that takes test requirements into account early in the design process.

die see chip. (The plural of die is dice.)

diffusion a thermochemical process whereby controlled impurities are introduced into the silicon to define the piezoresistor. Compared to ion implantation, it has two major disadvantages: (1) the maximum impurity concentration occurs at the surface of the silicon, rendering it subject to surface contamination, and making it nearly impossible to produce buried piezoresistors; (2) control over impurity concentrations and levels is about one thousand times poorer than obtained with ion implantation.

digital output transducer output that represents the magnitude of the measurand in the form of a series of discrete quantities coded in a system of notation.

DIN Deutsches Industrial Norm; a set of technical/scientific and dimensional standards developed by a German organization.

direct sequence fixed-frequency spread-spectrum technique.

dissipation constant the dissipation constant is the ratio, (in milliwatts per °C) at a specified ambient temperature, of a change in power dissipation in a thermistor to the resultant body temperature change.

DLC data link controller.

Doppler effect the change in the observed frequency of a wave caused by a time rate of change in the distance between the source and point of observation.

Doppler Radar radar that exploits the Doppler effect to measure the radial component of the relative velocity between the radar system and the target.

downconverter a device that provides gain and frequency translation to a lower frequency.

DPSK differential phase shift keying; modulation technique for transmission where the frequency remains constant but phase changes from 90°, 180°, and 270° to define the digital information.

drift an undesired change in output over a period of time with constant input applied.

DSP digital signal processing; a process by which a sampled and digitized data stream is modified in order to extract relevant information. Also, a digital signal processor.

dynamic range the ratio of the largest to the smallest values of a range, often expressed in decibels.

EDA electronic design automation.

EDP ethylene diamine pyrocatechol; an etchant for bulk micromachining.

embedded control see embedded system.

embedded sensing the use of a sensor in a product for a secondary instead of a primary function (e.g., protection in a power device).

embedded system a system with one or more computational devices (which may be microprocessors or microcontrollers) that are not directly accessible to the user of the system.

end-point straight-line fit the maximum deviation of any data point on a sensor output curve from a straight line drawn between the end data points on the output curve.

EEPROM electrically erasable programmable read only memory; a semiconductor technique used for permanent storage. It can be reprogrammed in the system.

EMI electromagnetic interference; see RFI.

engine the computational portion of an IC.

epitaxial or epi a single-crystal semiconductor layer grown upon a single-crystal substrate and having the same crystallographic characteristics as the substrate material.

EPROM erasable programmable read only memory; a semiconductor technique used for permanent storage, but can be erased by ultraviolet light.

error the algebraic difference between the indicated value and the true value of the input. Error is usually expressed in percent of full-scale span and sometimes expressed in percent of the sensor output reading.

error band the band of maximum deviations of the output values from a specified reference line or curve due to those causes attributable to the sensor. Usually expressed as ''+ % of full-scale output.'' The error band should be specified as applicable over at least two calibration cycles, so as to include repeatability, and verified accordingly.

ESD electrostatic discharge; an electrical discharge usually of high voltage and low current.

ESDA electronic systems design automation.

etch stop a layer of P-type material that etches at a much slower rate than N-type material.

Ethernet local area network software standard that determines system operation.

ETSI European Telecommunications Standards Institute.

excitation voltage (current) (also see supply voltage (current)) the external electrical voltage and/or current applied to a sensor for its proper operation (often referred to as the supply circuit or voltage).

FCC Federal Communications Commission.

FDMA frequency-division multiple access; each active user in multiple user system is assigned an individual frequency channel.

FEA finite element analysis; mechanical simulation of stress.

FFT fast Fourier transform; algorithm for reducing the number of calculations to extract frequency information from a signal.

Fieldbus a two-way communications link among intelligent field devices and control systems devices.

FIR finite impulse response; filter whose output is determined by its coefficients and previous inputs. It is characterized by a linear phase response.

firmware computer program or instructions stored in ROM instead of in the software.

filter a circuit that reduces noise and other unwanted elements of a signal.

flash semiconductor memory that can be used for permanent storage and is easily electrically reprogrammed in the system; faster than EEPROM.

floating point representation of numbers in scientific notation with the exponent and the mantissa given separately to accommodate a very wide dynamic range.

flops floating point operations per second; a measurement of microprocessor performance.

FM frequency modulation; carrier wave modulation using frequency variation in proportion to the amplitude of the modulating signal.

FPGA field-programmable gate array; an IC that can be programmed after it is assembled into a circuit.

frame a group of bits sent serially over a communications channel. The logical transmission unit sent between data link layer entities contains its own control information for addressing and error checking.

frequency hopping in spread-spectrum approach, both base and subscriber (or handset and base) hop from frequency to frequency simultaneously to minimize data lost through interference.

FSK frequency shift keying; a method of frequency modulation in which the modulating wave shifts the output frequency between predetermined values, and the output wave has no phase discontinuity.

full-scale output the output at full-scale measurand input at a specified supply voltage. This signal is the sum of the offset signal plus the full-scale span.

full-scale span the output variation between zero and when the maximum recommended operating value is applied.

fuzzy logic a branch of logic that uses degree of membership in sets rather than a strict true/false membership.

GaAs gallium arsenide; a compound semiconductor material that has higher speed and higher operating temperature capability than silicon.

gateway a node used to connect networks that use different protocols: a protocol converter.

genetic algorithms guided stochastic search techniques that utilize the basic principles of natural selection to optimize a given function.

GMSK Gaussian minimum shift keying; a modulation technique for RF signal transmission.

GPS global positioning system.

Harvard architecture on-chip program and data are in two separate spaces and are carried in parallel by two separate buses.

HDL hardware description language.

HLL high-level language; a programming language that utilizes macro statements and instructions that closely resemble human language or mathematical notation to describe the problem to be solved or the procedure to be used.

hysteresis also see temperature hysteresis hysteresis refers to a transducer's ability to reproduce the same output for the same input, regardless of whether the input is increasing or decreasing.

IC integrated circuit; a semiconductor that has several hundred or more transistors designed into it.

IEEE Institute of Electrical and Electronics Engineers.

IIR infinite impulse response; filter whose output is determined by its coefficients, current output, and previous inputs. It is characterized by a nonlinear phase response.

implant see ion implantation.

infrared area in the electromagnetic spectrum ranging from 1 to 1,000 microns.

input impedance (resistance) the impedance (resistance) measured between the positive and negative (ground) input terminals at a specified frequency with the output terminals open.

instrument a system that allows measured quantity to be observed as data. An instrument consists of a sensing element and conversion, manipulation, and data transmission and presentation elements. All elements are not necessarily present in a particular instrument.

integrated the combined design and fabrication of interconnected components. An integrated circuit contains a multiplicity of transistors as well as diodes, resistors, sensor(s), and so forth.

integration the combination of previously discrete or separate circuit designs.

interchangeability the ability of multiple systems of devices to exchange information and mutually use the information that has been exchanged between systems and devices.

interoperability the capability of a computing or control system to replace component parts or devices with component parts or devices of different manufacturers and different product families while maintaining the full range (or partial range) of optimal system functionality.

I/O input/output for a semiconductor or circuit.

ion beam milling a dry-etching process that uses an ion beam to remove material through a sputtering action.

ion implantation a process whereby impurity ions are accelerated to a specific energy level and impinged upon the silicon wafer. The energy level determines the depth to which the impurity ions penetrate the silicon. Impingement time determines the impurity concentration.

ISA Instrument Society of America.

ISM industrial, scientific, and medical frequency bands that were allocated by the FCC to spur rapid development of RF applications in a virtual open-market fashion. Licensing is automatic.

ISO International Standards Organization.

isotropic etching that is independent of crystallographic orientation.

ITS intelligent transportation system.

IVHS intelligent vehicle-highway system, now ITS (intelligent transportation system).

JEDEC Joint Electron Devices Engineering Council.

JIS Joint Industrial Standards. Also Japanese Industrial Standards Committee (JISC).

kernel software that is embedded in the controller.

KGD known good die; specially tested die.

laser trimming (automated) a method for adjusting the value of thin- or thick-film resistors using a computer-controlled laser system.

latency guaranteed access (with maximum priority) within a defined time.

Lidar laser infrared radar.

LIGA process developed by the Fraunhofer Institute in Munich that combines X-ray lithography, electroforming, and molding to obtain high-aspect ratio micromachined structures. LIGA is derived from the German terms for lithography, electroforming, and molding.

linearity error the maximum deviation of the output from a straight-line relationship input over the operating measurand range. The type of straight-line relationship (end point, least squares approximation, etc.) should be specified.

load impedance the impedance presented to the output terminals of a sensor by the associated circuitry.

Loran C third generation long-range navigation system that establishes position by measuring the time differences of signals arriving from fixed location transmitters.

Manchester a digital signaling technique in which there is a transition in the middle of each bit time. A "1" is encoded with a high level during the first half of the bit time and a "0" is encoded with a low level during the second half of the bit time.

maximum operating temperature the maximum body temperature at which the sensor will operate for an extended period of time with acceptable stability of its characteristics. This temperature is the result of the internal or external heating, or both, and should not exceed the maximum value specified.

maximum power rating the maximum power that a sensor will dissipate for an extended period of time with acceptable stability of its characteristics.

MCM multichip module; the interconnection of two or more semiconductor chips in a semiconductor-type package.

MCU microcontrol unit, or microcontroller unit; a semiconductor that has a CPU, memory, oscillator, and I/O capability on the same chip.

mechatronics the synergistic combination of precision mechanical engineering, electronic control, and the systems approach for designing products and manufacturing processes.

media means of data transmission in a network (e.g., two-wire, twisted-pair, coaxial cable, power lines, single-wire (with common ground), infrared, radio frequency, fiber-optic, etc.).

MEMS microelectromechanical systems.

MESFET metal semiconductor field effect transistor; a high-frequency semiconductor device produced in GaAs semiconductor technology.

microcode machine instructions built permanently into controller circuitry.

microcontroller MCU; a single integrated circuit that contains a CPU, memory, a clock oscillator, and I/O.

microengine the computational portion of an IC.

micromachining the chemical etching of mechanical structures in silicon or other semiconductor material, usually to produce a sensor or actuator.

micron 10^{-6} meter.

MIPS millions of instructions per second; a measurement of microprocessor throughput.

mixed signal the combination of analog and digital circuitry in a single semiconductor process.

mixer device that utilizes nonlinear characteristics to provide frequency conversions from one frequency to another.

MMIC monolithic microwave integrated circuit; a high-frequency integrated circuit.

modem modulator-demodulator; unit that modulates and demodulates digital information from a terminal or computer port to an analog carrier signal for passage over an analog line.

monotonicity a measurement of linearity. A monotonic curve is one in which the dependent variable either always increases or decreases as the independent variable.

MPU microprocessor unit; a central processing unit on a chip, usually without input/output or primary memory storage.

multiplex (MUX) the combining of several messages for transmission over the same signal path.

multiplexer device that allows two or more signals to be transmitted simultaneously on a single carrier or channel.

multiprocessing the simultaneous execution of two or more instructions by a computer.

nanometer 10^{-9} meter.

negative temperature coefficient (NTC) an NTC thermistor is one which the zero-power resistance decreases with an increase in temperature.

neural network a collection of independent processing nodes that communicate with one another in a manner analogous to the human brain.

Neuron™ a VLSI component that performs the network and application-specific processing within a node.

NIST National Institute of Standards and Technology in Gaithersburg, MD. Formerly, the National Bureau of Standards.

NMOS n-channel metal oxide semiconductor.

node any subassembly of a multiplex system that communicates on the signal bus.

NRZ nonreturn to zero; a data format in which the voltage or current value (typically voltage) determines the data bit value (one or zero).

null offset the electrical output present when the sensor is at null.

null temperature shift the change in null output value due to a change in temperature.

null output see zero offset.

Nyquist rate lowest sampling rate necessary to completely reconstruct a signal without distortion due to aliasing. Equal to twice the highest frequency component in the signal.

offset see zero offset.

OOK on-off keying; the simplest form of ASK, in which the carrier is switched on and off by the pulse code modulation waveform.

open loop system with no sensory feedback.

operating temperature range the range of temperature between minimum and maximum temperature at which the output will meet the specified operating characteristics.

operating impedance the impedance measured between the positive and negative (ground) output terminals at a specific frequency with the input open.

OSI model the open system interconnect model defines a seven-layer model for a data communications network.

parallel computing the processing of two or more (often many more) programs at one time by interconnected processing systems.

partitioning system design methodology that determines which portion of the circuit is integrated using a particular silicon process instead of completely integrating design using one process.

PCMCIA Personal Computer Memory Card International Association.

peripherals external circuit components necessary to achieve desired functionality, usually for an MCU or MPU.

photolithography process in which a pattern in a mask is transferred to a wafer, resulting in areas to be doped or selectively removed.

PID proportional integral derivative; control technique commonly used in servo systems.

piezoelectric effect the ability of certain materials to become electrically polarized in response to applied strain or to be strained in response to applied voltage.

piezoresistor a resistor that changes resistance in response to applied strain.

pipeline bus structure within an MPU that allows concurrent operations to occur.

plasma etching an etching process that uses an etching gas instead of a liquid to chemically etch a structure.

PLC programmable logic control.

PLL phase locked loop; a major component in a frequency synthesizer.

polysilicon (poly) silicon composed of randomly arranged crystal unit cells.

positive temperature coefficient (PTC) a PTC thermistor is one which the zero-power resistance increases with an increase in temperature.

precision the smallest discernible change in the output signal of a device.

protocol the rules governing the exchange of data between networked elements.

QPSK quadrature phase shift keying.

radar radio detection and ranging; system based on transmitted and reflected RF energy for determining and locating objects, measuring distance and altitude, and navigating.

ratiometric (ratiometricity error) at a given voltage, sensor output is a proportional value of that supply voltage. Ratiometricity error is the change in this proportion resulting from any change to the supply voltage. (Usually expressed as a percent of full-scale output.)

range see operating (temperature) range.

RAM random access memory; memory used for temporary storage of data. This is volatile memory that is lost when the power is turned off.

reactive ion etching (RIE) a dry-etching process that combines plasma etching and ion beam removal of the surface layer.

real-time system system that accepts external inputs, processes them, and produces outputs in a fixed period of time.

repeatability the maximum change in output under fixed operating conditions over a specified period of time.

resistance ratio characteristic the ratio of the zero-power resistance of a thermistor measured at 25°C to that resistance measured at 125°C.

resistance temperature characteristic the relationship between the zero-power resistance of a thermistor and its body temperature.

resolution the maximum change in pressure required to give a specified change in output.

response time time required for a sensor output to change from its previous state to a final state within an error tolerance band of the new correct value.

RF radio frequency; frequencies of the electromagnetic spectrum normally associated with radiowave propagation. Sometimes defined as transmission of any frequency at which coherent electromagnetic energy radiation is possible, usually above 150 kHz.

RFI radio frequency interference; (usually) unintentionally radiated energy that may interfere with the operation of, or even damage, electronic equipment.

RISC reduced instruction set computer; a CPU architecture that optimizes processing speed by the use of a smaller number of basic machine instructions.

ROM read only memory; memory used for permanent storage of data and is nonvolatile memory.

router connects the channels in multiple-media control systems and passes information packets back and forth.

sacrificial layer a thin film deposited in the surface micromachining process that is later etched away to release a microstructure.

SAE Society of Automotive Engineers.

SAW surface acoustic wave.

scaleable ability of MCU or MPU architecture to be modified to meet the needs of several applications providing competitive price-performance points.

self-generating providing an output signal without applied excitation such as a thermo-electric transducer.

self-heating internal heating resulting from electrical energy dissipated within the unit.

sensactor a linguistic amalgam proposed by Ford engineers of a combined sensor, microprocessor, and actuator possibly integrated on a single silicon chip.

sensitivity the change in output per unit change in input for a specified supply or current.

sensitivity shift a change in sensitivity resulting from an environmental change, such as temperature.

sensor a device that provides a useful output to a specified measurand.

Sematech consortium of United States semiconductor and equipment manufacturers.

semiconductor sensor sensor manufactured using silicon, GaAs, SiC, or other semiconductor materials.

SIA Semiconductor Industry Association.

SiC silicon carbide; a high-temperature semiconductor, sensor, and MEMS material.

simulation design approach using computer models to predict circuit or system performance.

silicon compiler a tool that translates algorithms into a design layout for silicon.

silicon fusion bonding a process for bonding two silicon wafers at the atomic level without applying glue or an electric field, also called direct wafer bonding.

skew rate difference in delay between parallel paths.

slew rate maximum rate of change of voltage with time.

SMD surface mount device; see SMT.

smart sensor a device with built-in intelligence, whether apparent to the user or not (IEEE-proposed definition).

smart power (ICs) hybrid or monolithic devices that are capable of being conduction cooled, perform signal conditioning, and include a power control function such as fault management and/or diagnostics. The scope (of this definition) shall apply to devices with a power dissipation of 2W or more, capable of operating at a case temperature of 100°C and with a continuous current of 1A or more (JEDEC definition).

SMT surface mount technology; method of attaching components, both electrically and mechanically, to the surface of a conductive pattern.

span the output voltage variation given between zero and any given measurand input.

SPECfp92 floating point benchmark test and rating (1992) for comparing microprocessor computing power.

SPECint92 integer benchmark test and rating (1992) for comparing microprocessor computing power.

spectral filter a filter that restricts the electromagnetic spectrum to a specific bandwidth.

spin-on semiconductor process to deposit films and coatings.

spread spectrum technique used to reduce and avoid interference by taking advantage of statistical means to send a signal between two points. A figure of merit for spread-spectrum systems is "spreading gain" measured in dB. The two types of commercial spread-spectrum techniques are frequency hopping and direct sequence.

sputter semiconductor process to deposit a thin film of material, typically a metal, on the surface.

squeeze-film damping effect of ambient gases and spacing on the movement of micro-machined structures.

SSVS super smart vehicle systems; term used in Japan for vehicles with several new electronic systems, typically used in ITS.

stability the ability of a sensor to retain specified characteristics after being subjected to designated environmental or electrical test conditions.

state machine logic circuitry that, when clocked, sequences through logical operations and can be a preprogrammed set of instructions or logic states.

stiction static friction; adhering of surface micromachined layers due to capillary forces generated during wet-etching.

storage temperature range the range of temperature between minimum and maximum that can be applied without causing the sensor (unit) to fail to meet the specified operating characteristics.

submicron measurement of the geometries or critical spacing used for complex, highly integrated circuits.

superscalar the ability of an MPU to dispatch multiple instructions per clock from a conventional linear instruction stream.

supply voltage (current) the voltage (current) applied to the positive and negative (ground) terminals.

surface micromachining a process for depositing and etching multiple layers of sacrificial and structural thin films to build complex microstructures.

TAB tape automated bonding; semiconductor packaging technique that uses a tiny lead-frame to connect circuitry on the surface of the chip to a substrate instead of wire bonds.

TDMA time-division multiple access; a technique that assigns each subscriber desiring service a different time slot on a given frequency. Signal compression is achieved by running at very high frequencies. Each user can then deliver the fixed packet message in a brief burst of time, thereby increasing the capacity of the system.

telemetry a remote measurement that permits the data to be interpreted at a distance from the detector.

temperature coefficient of full-scale span the percent change in the full-scale span per unit change in temperature relative to the full-scale span at a specified temperature.

temperature coefficient of resistance the percent change in the dc input impedance per unit change in temperature relative to the dc input impedance at a specified temperature.

temperature error the maximum change in output at any input value in the operating range when the temperature is changed over a specified temperature range.

temperature hysteresis the difference in output at any temperature in the operating temperature range when the temperature is approached from the minimum operating temperature and when approached from the maximum operating temperature with zero input applied.

temperature-wattage characteristic the relationship, at a specified ambient temperature, between the thermistor temperature and the applied steady-state wattage.

thermal compensation see compensation.

thermal offset shift see temperature coefficient of offset.

thermal span shift see temperature coefficient of full-scale span.

thermal zero shift see temperature coefficient of offset.

thermal time constant the time required for a thermistor to change to 63.2 percent of the total difference between its initial and final body temperature when subjected to a step function change in temperature under zero-power conditions.

thermistor a thermally sensitive resistor whose primary function is to exhibit a change in electrical resistance with a change in body temperature.

thin film a technique using vacuum deposition of conductors and dielectric materials onto a substrate (frequently silicon) to form an electrical circuit.

torr pressure unit equivalent to 1 mm of mercury (mmHg).

transceiver the combination of radio transmitter and receiver, usually using some common circuitry, in a portable or mobile application.

transducer a device converting energy from one domain into another and calibrated to minimize the errors in the conversion process.

transfer function a mathematical, graphical, or tabular statement of the influence that a system or element has on the output compared at the input and output terminals.

transponder a transmitter-receiver that transmits signals automatically when a triggering signal is received.

tribology the study of friction, wear, and lubrication in surfaces sliding against each other, as in bearings and gears.

standard reference temperature the thermistor body temperature at which nominal zero-power resistance is specified (e.g., 25°C).

ULSI ultra large scale integration; a chip with over 1,000,000 components.

upconverter an upmixer with amplification, and possibly other functions such as power control and/or transmit envelope shaping.

VCO voltage-controlled oscillator; an oscillator whose output frequency varies with an applied dc control voltage.

via vertical length of metal deposited through small hole in oxide used to electrically connect two layers of metal.

VLSI very large scale integration; a chip with 100,000 to 1,000,000 components.

Verilog a hardware description language for behavior-level circuit design placed in the public domain by Cadence Design Systems, Inc.

VHDL VHSIC hardware description language; a hardware description language for behavior level circuit design developed by the U.S. Department of Defense.

VHSIC very high speed integrated circuit.

Von Neumann (architecture) program and data are carried sequentially on the same bus.

wafer a thin, usually round slice of semiconductor material, from which chips are made.

X-ray (photo)lithography technique use to achieved small physical spacing in integrated circuits.

zero offset the output at zero input for a specified supply voltage or current.

About the Author

Randy Frank is the Technical Marketing Manager for Motorola's Semiconductor Products Sector in Phoenix, Arizona. He is also an Application Specialist in sensor technology. He has a BSEE, an MSEE, and an MBA from Wayne State University in Detroit, Michigan, and over 25 years experience in automotive and control systems engineering, including the development of electronic engine controls and smart-power semiconductor products. Mr. Frank has been the chairman and is currently a member of the Sensor Standards Committee of SAE (Society of Automotive Engineers). He is also a member of the IEEE Sensor Terminology Taskforce and a former member of the A.A.M.I. (Association for the Advancement of Medical Instrumentation) Blood Pressure Transducer Committee. He has taught a course in Advanced Instrumentation and Control at the University of Michigan, which included the reliability aspects of sensors. He has three patents issued and has published over 150 technical papers and articles and several book chapters on semiconductor products and applications.

Index